Understanding the Long-Term Evolution of the Coupled Natural-Human Coastal System

THE FUTURE OF THE U.S. GULF COAST

Committee on Long-Term Coastal Zone Dynamics: Interactions and Feedbacks Between Natural and Human Processes Along the U.S. Gulf Coast

Board on Earth Sciences and Resources
Ocean Studies Board
Division on Earth and Life Studies

Board on Environmental Change and Society
Division of Behavioral and Social Sciences and Education

A Consensus Study Report of

The National Academies of
SCIENCES · ENGINEERING · MEDICINE

THE NATIONAL ACADEMIES PRESS
Washington, DC
www.nap.edu

THE NATIONAL ACADEMIES PRESS • 500 Fifth Street, NW • Washington, DC 20001

This activity was supported by the Gulf Research Program. Any opinions, findings, conclusions, or recommendations expressed in this publication do not necessarily reflect the views of any organization or agency that provided support for the project.

International Standard Book Number-13: 978-0-309-47584-6
International Standard Book Number-10: 0-309-47584-8
Digital Object Identifier: https://doi.org/10.17226/25108

Additional copies of this publication are available for sale from the National Academies Press, 500 Fifth Street, NW, Keck 360, Washington, DC 20001; (800) 624-6242 or (202) 334-3313; http://www.nap.edu.

Printed in the United States of America

Suggested citation: National Academies of Sciences, Engineering, and Medicine. 2018. *Understanding the Long-Term Evolution of the Coupled Natural-Human Coastal System: The Future of the U.S. Gulf Coast:* Washington, DC: The National Academies Press. doi: https://doi.org/10.17226/25108.

The National Academies of
SCIENCES · ENGINEERING · MEDICINE

Consensus Study Reports published by the National Academies of Sciences, Engineering, and Medicine document the evidence-based consensus on the study's statement of task by an authoring committee of experts. Reports typically include findings, conclusions, and recommendations based on information gathered by the committee and the committee's deliberations. Each report has been subjected to a rigorous and independent peer-review process and it represents the position of the National Academies on the statement of task.

Proceedings published by the National Academies of Sciences, Engineering, and Medicine chronicle the presentations and discussions at a workshop, symposium, or other event convened by the National Academies. The statements and opinions contained in proceedings are those of the participants and are not endorsed by other participants, the planning committee, or the National Academies.

For information about other products and activities of the National Academies, please visit www.nationalacademies.org/about/whatwedo.

COMMITTEE ON LONG-TERM COASTAL ZONE DYNAMICS: INTERACTIONS AND FEEDBACKS BETWEEN NATURAL AND HUMAN PROCESSES ALONG THE U.S. GULF COAST

TUBA ÖZKAN-HALLER (*Chair*), Oregon State University
GREGORY A. CARTER, University of Southern Mississippi
JUST CEBRIAN, Dauphin Island Sea Lab and University of South Alabama
ROBERT (TONY) A. DALRYMPLE (NAE[1]), Johns Hopkins University
JORDAN R. FISCHBACH, Water and Climate Resilience Center and RAND Corporation
JENNIFER L. IRISH, Virginia Tech
ALEXANDER S. KOLKER, Louisiana Universities Marine Consortium and Tulane University
SHUBHRA MISRA, Chevron Energy Technology Company (until May 2018)
LAURA J. MOORE, University of North Carolina at Chapel Hill
MARTIN D. SMITH, Duke University
TORBJÖRN E. TÖRNQVIST, Tulane University
GABRIELLE WONG-PARODI, Carnegie Mellon University

National Academies of Sciences, Engineering, and Medicine Staff

DEBORAH GLICKSON, Senior Program Officer
HEATHER KREIDLER, Associate Program Officer
COURTNEY DEVANE, Administrative Coordinator
JAMES HEISS, Postdoctoral Fellow

[1] National Academy of Engineering.

BOARD ON EARTH SCIENCES AND RESOURCES

GENE WHITNEY (*Chair*), Congressional Research Service (Retired)
R. LYNDON ARSCOTT (NAE[1]), International Association of Oil & Gas Producers (Retired)
BRENDA B. BOWEN, The University of Utah
CHRISTOPHER CAMERON, GeoLogical Consulting, LLC
RODNEY C. EWING, Freeman Spogli Institute for International Studies and Stanford University
CAROL P. HARDEN, Department of Geography, University of Tennessee
THORNE LAY (NAS[2]), Department of Earth and Planetary Sciences, University of California, Santa Cruz
ANN S. MAEST, Buka Environmental
ZELMA MAINE-JACKSON, Washington State Department of Ecology, Nuclear Waste Program
MARTIN W. McCANN, Jack R. Benjamin and Associates and Stanford University
JAMES M. ROBERTSON, Wisconsin Geological and Natural History Survey (Retired)
JEFFREY N. RUBIN, Tualatin Valley Fire and Rescue
JAMES A. SLUTZ, National Petroleum Council
SHAOWEN WANG, University of Illinois at Urbana-Champaign
ELIZABETH J. WILSON, Dartmouth College

National Academies of Sciences, Engineering, and Medicine Staff

ELIZABETH EIDE, Director
ANNE LINN, Scholar
DEBORAH GLICKSON, Senior Program Officer
SAMMANTHA MAGSINO, Senior Program Officer
COURTNEY DEVANE, Administrative Coordinator
NICHOLAS ROGERS, Financial and Research Associate
YASMIN ROMITTI, Research Associate
CARLY BRODY, Senior Program Assistant
RAYMOND (REMY) CHAPPETTA, Senior Program Assistant
ERIC EDKIN, Senior Program Assistant

[1] National Academy of Engineering.
[2] National Academy of Sciences.

[1] National Academy of Sciences.
[2] National Academy of Engineering.

BOARD ON ENVIRONMENTAL CHANGE AND SOCIETY

KRISTIE L. EBI (*Chair*), University of Washington
JOSEPH L. ARVAI, University of Michigan
HALLIE C. EAKIN, Arizona State University
LORI M. HUNTER, University of Colorado Boulder
KATHARINE L. JACOBS, Center for Climate Adaptation Science and Solutions and University of Arizona
MICHAEL ANTHONY MENDEZ, Yale University
RICHARD G. NEWELL, Resources for the Future
MARY D. NICHOLS, California Air Resources Board
JONATHAN T. OVERPECK, University of Michigan
ASEEM PRAKASH, University of Washington
MAXINE L. SAVITZ, Honeywell, Inc. (Retired)
MICHAEL P. VANDENBERGH, Vanderbilt University Law School
JALONNE L. WHITE-NEWSOME, The Kresge Foundation
ROBYN S. WILSON, The Ohio State University

National Academies of Sciences, Engineering, and Medicine Staff

TOBY WARDEN, Board Director
HEATHER KREIDLER, Associate Program Officer
JORDYN WHITE, Program Officer
TINA M. LATIMER, Program Coordinator
LETICIA GARCILAZO GREEN, Senior Program Assistant

Acknowledgments

This Consensus Study Report was greatly enhanced by the participants of meetings held as part of this study. The committee acknowledges those who gave presentations at or participated in committee meetings: Damarys Acevedo-Mackey (U.S. Army Engineer Research and Development Center), Mead Allison (The Water Institute of the Gulf), Phil Berke (Texas A&M University), Carl Bernier (Rice University), Cas Bridge (Chevron), Joe Calantoni (Naval Research Laboratory), Noreen Clancy (RAND Corporation), Craig Colten (Louisiana State University), Brady Couvillion (U.S. Geological Survey), Don Danmeier (Chevron), David Dismukes (Louisiana State University), Tom Drake (Office of Naval Research), Nicole Elko (American Shore and Beach Preservation Association), Carl Ferraro (Alabama Department of Conservation and Natural Resources), Angelina Freeman (Louisiana Coastal Protection and Restoration Authority), Mathew Hauer (University of Georgia), David Kidwell (National Oceanic and Atmospheric Administration), Maria Lemos (University of Michigan), Kelli Levy (Environmental Management Division Director, Pinellas County), Tucker Mahoney (Federal Emergency Management Agency), John Alex McCorquodale (University of New Orleans), Ehab Meselhe (The Water Institute of the Gulf), Shahzaad Mohammed (Cheniere LNG), Ryan Moyer (Florida Department of Environmental Protection), Brad Murray (Duke University), Ray Newby (Texas General Land Office), Natalie Peyronnin (Environmental Defense Fund), George Ramseur (Mississippi Department of Marine Resources), Denise Reed (The Water Institute of the Gulf), Hugh Roberts (ARCADIS), Dano Roelvink (IHE Delft Institute for Water Education), Julie Dean Rosati (U.S. Army Corps of Engineers, Coastal and Hydraulics Laboratory), Jim Schock (Florida Building Commission), Hilary Stockdon (U.S. Geological Survey), Ariana Sutton-Grier (University of Maryland), LaDon Swann (Mississippi-Alabama Sea Grant Consortium), Sherri Swanson (HDR Engineering, Inc.), Jason Theriot (historian and author), Elizabeth Vargas (Texas General Land Office), Jenneke Visser (University of Louisiana at Lafayette), Ian Voparil (Shell Deepwater Gulf of Mexico), George Voulgaris

(National Science Foundation), Ty Wamsley (U.S. Army Corps of Engineers), and Eric White (The Water Institute of the Gulf).

This Consensus Study Report was reviewed in draft form by individuals chosen for their diverse perspectives and technical expertise. The purpose of this independent review is to provide candid and critical comments that will assist the National Academies of Sciences, Engineering, and Medicine in making each published report as sound as possible and to ensure that it meets the institutional standards for quality, objectivity, evidence, and responsiveness to the study charge. The review comments and draft manuscript remain confidential to protect the integrity of the deliberative process.

We thank the following individuals for their review of this report:

John Anderson, Rice University
R. Lyndon Arscott, International Association of Oil & Gas Producers (Retired)
Craig Colten, Louisiana State University
Jamie Kruse, East Carolina University
Nirnimesh Kumar, University of Washington
Maria Carmen Lemos, University of Michigan
Allen Marr, Geocomp Corporation
Bonnie McCay (NAS[1]), Rutgers University
Richard McLaughlin, Texas A&M University–Corpus Christi
Denise Reed, University of New Orleans
Mark Stacey, University of California, Berkeley
Ariana Sutton-Grier, The Nature Conservancy
Robert Weisberg, University of South Florida

Although the reviewers provided many constructive comments and suggestions, they were not asked to endorse the conclusions or recommendations of this report nor did they see the final draft before its release. The review of this report was overseen by **Michael C. Kavanaugh** (NAE[2]), Geosyntec Consultants, and **Linda K. Blum,** University of Virginia. They were responsible for making certain that an independent examination of this report was carried out in accordance with the standards of the National Academies and that all review comments were carefully considered. Responsibility for the final content rests entirely with the authoring committee and the National Academies.

[1] National Academy of Sciences.
[2] National Academy of Engineering.

Contents

Summary

The U.S. Gulf of Mexico Coast (hereafter referred to as "the Gulf Coast") provides a valuable setting to study deeply connected natural and human interactions and feedbacks that have led to a complex, interconnected coastal system. The physical landscape in the region has changed significantly due to broad-scale, long-term processes such as coastal subsidence and river sediment deposition, as well as short-term episodic events such as hurricanes. Modifications from human activities, including building levees and canals and constructing buildings and roads, have left their own imprint on the natural landscape. Indeed, part of the Gulf Coast's uniqueness is the concentration of a wide range of energy-related infrastructure in the region. This coupled natural-human coastal system and the individual aspects within it (physical, ecological, and human) are under increased pressure from accelerating environmental stressors such as sea level rise, intensifying hurricanes, and continued population increase with its accompanying coastal development. Promoting the resilience and maintaining the habitability of the Gulf Coast into the future will need improved understanding of the coupled natural-human coastal system, as well as effective sharing of this understanding in support of decision making and policies.

The National Academies of Sciences, Engineering, and Medicine's (the National Academies') Gulf Research Program (GRP) asked the National Academies' Board on Earth Sciences and Resources, Ocean Studies Board, and Board on Environmental Change and Society to undertake a study on long-term coastal zone dynamics along the Gulf Coast. The ad hoc committee was asked to identify scientific and technical gaps in understanding the interactions and feedbacks between human and natural processes, to define essential components of a research and development program in response to the identified gaps, and to develop and set priorities for up to three critical areas of research. In addition, the committee was asked to identify barriers to and opportunities for more effective communication among scientists and coastal stakeholders about long-term changes to the coastal zone (see Box 1.1 for the full Statement of Task). The project originated as a way to better understand the multiple and interconnected factors that influence long-term processes along the Gulf Coast and was envisioned to help inform decision making and research planning related to the strategic initiatives of the GRP.

In this report, the committee identifies three critical areas of research that encompass high-priority gaps in scientific knowledge that, if addressed, will increase understanding of the coupled natural-human system along the Gulf Coast. The committee also presents a research agenda that should be undertaken to meet these gaps, as well as strategies that could guide such an agenda. Finally, barriers to effective communication between scientists and stakeholders, as well as opportunities to move past these barriers, are considered.

To assist with these tasks, the committee convened three information-gathering meetings, including a workshop with 20 invited participants; heard presentations from leaders in related fields, including state and federal agencies, academia, nongovernmental organizations, and the energy industry; and consulted peer-reviewed research literature, community-sponsored efforts, and state and federal government reports from the Gulf Coast and other relevant coastal regions.

CRITICAL RESEARCH AREAS

Better understanding of long-term coastal zone dynamics involves a thorough examination of the coupled natural-human coastal system. While the interconnected and complex nature of the system makes it difficult to discuss any one aspect of the Gulf Coast region in isolation, the system can most readily be grouped into three related areas: the physical and ecological components of the natural system, the human system, and interactions and feedbacks between the natural and human systems. The Gulf Coast is governed by a combination of physical drivers originating in the ocean, in the atmosphere, and on land. These include, among others, sea level rise, subsidence, hurricane and flooding hazards, and coastal morphology. Ecosystem dynamics, structure, and function are important aspects of the ecological system. The human system includes both coastal development and adaptive responses to coastal change, such as individual and community-level decisions on targeted investment, relocation, or migration.

The committee was asked to identify up to three critical areas of research that would increase understanding of long-term natural coastal dynamics in order to advance the science and help inform stakeholder decision making. First, the committee developed a comprehensive vision that could guide the critical areas of research *to understand and predict the feedbacks and interactions among the physical, ecological, and human components and the resulting evolution of the coupled system along the Gulf Coast, in the context of both human and climate drivers*. Then, the committee focused on the relevant timescales for this vision: a near-decadal scale (10–50 years) and a decadal-century scale (50–200 years). These periods encompass both the timescales of the physical and ecological drivers of anticipated changes and the motivating factors for human response and decision making. Next, the three critical areas for research were identified.

- **Critical Area 1:** How will coastal landforms and coastal ecosystems along the Gulf Coast respond to rapidly changing conditions (both natural and human induced), especially given the expectation for continued relative sea level rise acceleration?
- **Critical Area 2:** How will human settlement and economic activity along the Gulf Coast respond to evolving coastal landforms and ecosystems under rapidly changing conditions?
- **Critical Area 3:** How can improved understanding of both near- and long-term evolution of the Gulf Coast coupled natural-human system be applied to inform stakeholder decisions made at local, state, and regional scales? How does the coupled natural-human system evolve when decision making is updated as scientific understanding advances?

RESEARCH GAPS

These critical areas encompass 12 gaps in the current scientific understanding of the coupled natural-human coastal system. These high-priority gaps, if addressed, will transform the present scientific understanding of the Gulf Coast coupled natural-human system and the ability to assess its future evolution. Research gaps were first identified through the consideration of one-way interactions among the various system components. They were viewed through a disciplinary lens (whether physical, ecological, or human) that could best enable progress to address each identified gap. Subsequently, the research gaps that require the consideration of feedbacks among all components of the system were discussed. It is important to note, however, that all of the research gaps (whether physical, ecological, or human) are interrelated to some degree and are intended to be thought of as integral to the Gulf Coast coupled natural-human system.

The Natural System: Physical Processes

Physical processes that drive changes in the natural coastal system occur over varying timescales that range from episodic (e.g., hurricanes, rainfall events) to longer-term processes (e.g., rising sea level, subsidence). These processes also act across a range of spatial scales and cause short-term disturbances such as flooding, shoreline erosion, and changes in water quality, as well as longer-term changes in the landscape such as wetland loss and the migration of river channels, barrier islands, and tidal inlets. Understanding key physical processes of the system, and how they interact with each other, will lead to better ecosystem management and human decision making to deal with coastal change. The following gaps in understanding are associated with Critical Area 1 (how coastal landforms and ecosystems respond to changing conditions):

- **Research Gap 1:** Current datasets, monitoring systems, and approaches are insufficient to track and understand how the **oceanic component** of sea level (i.e., excluding subsidence) is changing along the Gulf Coast and to predict how it will change in the future.
- **Research Gap 2:** The causes, rates, and patterns of **subsidence** along the Gulf Coast are not sufficiently well understood to allow for accurate prediction at the local to regional scale.
- **Research Gap 3:** The **combined effects** of freshwater input from Gulf Coast watersheds, storm surge, sea level rise, and development on **coastal flood hazards** are not well understood, thereby limiting the capacity to include and model these effects in predictions of Gulf Coast dynamics.
- **Research Gap 4:** The relative contributions of naturally occurring and artificially managed **riverine sediment delivery** (e.g., availability, fluxes), diversion and management activities, and how they impact the **evolution of coastal landforms** (e.g., river deltas, barrier islands) and ecosystems (e.g., wetlands) is poorly understood.
- **Research Gap 5:** Limited understanding of **sediment transport processes** and uncertainties in predicting **future hydrodynamic conditions** hampers the ability to project long-term coastal evolution.
- **Research Gap 6:** There is a critical need to understand and project the **future response of coastal landforms** and embayments to **changing climate** and the conditions under which they will no longer be able to keep pace with relative sea level rise.

The Natural System: Ecological Processes

Gulf Coast ecosystems evolve continuously over decadal to centennial scales. Humans have imposed and will continue to impose substantial change from factors such as changes in the built environment, industrial activities, and climate change. Placing the ecological component within the Gulf Coast coupled natural-human system first entails understanding how ecosystems function under natural conditions, and then how human alterations affect those functions. Below are salient gaps in the understanding of ecosystem function and management under current and future conditions, also tied to Critical Area 1.

- **Research Gap 7:** There is limited understanding of the individual and combined effects of current **environmental gradients, physical forcing, climate change,**

and coastal development (including energy-related infrastructure) on Gulf Coast ecosystems.
- **Research Gap 8:** The understanding of **strategic natural resource conservation and restoration activities** for effective coastal management is limited.

The Human System

Understanding the evolution of the coupled natural-human coastal system necessitates better knowledge of the aspects of the human system that interact and feed back with the natural system. In addition, the ability to predict significant human processes with confidence over decadal to centennial timescales is poor and will be particularly challenged if thresholds of behavior, status, or perceptions are exceeded, leading to new equilibrium states for the coupled natural-human system components. The following research gaps fall within Critical Area 2 (the response and adaptation of Gulf Coast residents to changes in coastal landforms and ecosystems):

- **Research Gap 9:** There is a need to understand how **decisions about the built environment will be affected by coastal change** and how these decisions create feedbacks between the natural and human systems.
- **Research Gap 10:** There is a need for better understanding of **how coastal changes affect the built environment** and which aspects of the built environment are most vulnerable to coastal changes.
- **Research Gap 11:** There is an incomplete understanding of the **vulnerability of different Gulf Coast communities** to coastal dynamics, how coastal dynamics trigger **migration and relocation decisions,** and how these decisions create feedbacks to the natural system.

The Coupled Natural-Human Coastal System

Physical drivers can cause not only modifications such as coastal erosion and landform migration, but also ecological alterations such as wetland loss and displacement of biological communities. Such changes can trigger human responses that, in turn, lead to further changes in the physical and ecological systems. Human activities can also be primary drivers for change in the natural system. For instance, the impacts of human development on coastal ecosystem function or physical processes can generate feedbacks for coastal communities. The understanding of these feedbacks for the coupled natural-human system in the Gulf Coast is limited. In addition, forecasting the behavior of the coupled natural-human coastal system is challenging, particularly for longer timescales (e.g., decades to centuries). This research gap is linked to Critical Area 3 (how improved understanding of the coupled system can inform decision making):

- **Research Gap 12:** Understanding how **decisions** about the built environment and human migration will affect the coupled natural-human coastal system is limited and can be furthered **through integrated modeling.**

A RESEARCH AGENDA FOR THE FUTURE

The independent, science-based National Academies' Gulf Research Program, established in 2013 as part of legal settlements with the companies involved in the 2010 *Deepwater Horizon* oil spill, has grant-making capabilities that present a unique and timely opportunity for the creation of a sustained, holistic research program focused on the Gulf Coast coupled natural-human system. The Gulf Research Program could leverage its scope and autonomy to create an integrated research program that addresses critical research areas and gaps in a sustained way, as state and federal funding agencies would be challenged to support such a program due to uncertainty in their budgets from year to year. Such an effort, with coordination and integration across multiple disciplinary research streams, has the potential to positively transform living along the Gulf Coast and in coastal zones around the world. This research program can be most successful if it includes the following components:

Focus on interactions and feedbacks critical to the evolution of the coupled natural-human coastal system. While it is very useful to study specific aspects of the natural and human systems, the greatest overall benefit is likely to come from focused efforts on the processes and mechanisms that are ultimately important to interactions and feedbacks between and among the natural and human systems.

Support collaborative, multidisciplinary research teams. Many of the current gaps in understanding of the coupled natural-human coastal system are complex and are unlikely to be addressed by a single discipline; rather, collaborative research teams involving multiple disciplines across the natural and social sciences will be needed.

Encourage comprehensive, Gulf Coast-wide, integrated observational and modeling efforts. Coordinating and integrating observational and modeling efforts will significantly amplify the gains that would otherwise be achieved by observation or modeling alone. The integration of observational and modeling programs, ideally through an iterative design, facilitates the development of targeted and adaptive observational programs, as well as the continued development of models and improvements in model skill.

Offer research opportunities that are longitudinal and multi-decadal. A program that intentionally takes the long view has the power to transform understanding of coastal evolution and to revolutionize the ability to project coastal change in the face of uncertain future conditions. Longitudinal observational, experimental, and monitoring programs can facilitate synthesis efforts aimed at tracking the drivers of change, quantifying patterns, and identifying cascading impacts through the system.

Deliver easily accessible, regularly updated observational data and model results. Making data and model results publicly available (especially in real or near-real time) and archiving them in accessible databases will extend their utility beyond scientific research and assist managers, planners, other researchers, and decision makers with adapting and responding to changing environments.

Coordinate at a high level. Management of a research and development program that includes long-term efforts that are Gulf Coast-wide and longitudinal, contains highly integrated modeling and observational components, and emphasizes interactions and feedbacks between the natural and human coastal systems needs intentional, consistent, and careful administration.

BARRIERS TO AND OPPORTUNITIES FOR EFFECTIVE COMMUNICATION

Addressing the research gaps will substantially advance understanding of the Gulf Coast coupled natural-human system and help identify salient feedbacks between humans and their environment. Turning research products into actionable policies for a more resilient future Gulf entails effective communication and collaboration between stakeholders and scientists. Specific barriers that currently prevent effective communication, as well as opportunities to overcome these barriers, are provided in the following section. In this report, "stakeholder" generally refers to a practitioner involved in coastal issues (e.g., planning or adaptation), such as a city planner or emergency manager. "Boundary organizations" play an intermediary role among different disciplines, creating and sustaining meaningful links between knowledge producers and users, and seek to provide a neutral ground for science-based discussion. A "boundary spanner" is an individual working at the edge of different groups who serves to connect those groups with each other.

Stakeholder Perspective

Barrier 1. Financial constraints, information availability, time, and expertise represent a barrier to effective communication. These factors make it difficult for stakeholders to know about, obtain, find, work with, and interpret information and data in a way that allows them to incorporate science into decision making.

Opportunity 1. Targeted funding opportunities that would allow practitioners to obtain data and to hire staff with the expertise and dedicated time to interpret scientific information would facilitate the use and application of available scientific information by other stakeholders. Alternatively, or in addition, the development of a Gulf Coast-wide repository of scientific information managed by well-informed staff who can provide support for stakeholders would be a valuable resource that would help facilitate the incorporation of science into decision making.

Barrier 2. Many scientific products that are intended to help inform decision making are not tailored to stakeholders' specific needs. As a result, the applicability of these products (e.g., tools, data, information) is not clear to stakeholders, who are then less likely to use them for decision making. Furthermore, many scientific products are not accompanied by sufficient instructions or training on how, why, or when to apply the provided information to the decision-making process; the information thus may go unused or may be applied inappropriately. Additionally, some products may be seen by stakeholders as serving the interests of one group over another and thus may not be seen as appropriate for decision making.

Opportunity 2. When developing products that are intended to inform decision making, scientists should be encouraged to engage substantively with stakeholders from the development to the delivery stage. Such an approach can create scientific products that are more likely to be effective and immediately applicable and may help to allay concerns over whether data are serving some needs over others. To encourage stakeholder involvement, solicitations for research programs might include a requirement for substantive and early engagement. Boundary organizations assist in facilitating this type of engagement, and including incentives for their involvement would further improve communication. The degree to which scientific information is effectively used could be further improved by streamlining and guiding the process by which stakeholders identify and access the information they need. Development of an innovative catalog of products would improve the abilities of stakeholders to access and apply these tools in their decision-making activities. Combining this effort with staff support (see Opportunity 1) would facilitate this process.

Barrier 3. The size and complexity of the energy industry, as well as apparent limitations to information sharing, present a barrier to effective communication between the energy industry and other stakeholders.

Opportunity 3. Create an incentive structure that fosters information sharing between the energy industry and other stakeholders, as well as protocols for how to engage more effectively to facilitate information sharing. This process could be facilitated by a third party such as a boundary organization.

Barrier 4. Limited financial and human resources, logistical complexity, and difficulty in identifying all relevant stakeholders, as well as skepticism, lack of understanding, or lack of trust by one or both parties, can make it difficult for practitioner stakeholders to communicate effectively with members of the general public, including vulnerable populations.

Opportunity 4. Boundary organizations can play a key role in facilitating trusting relationships among community members, practitioners, and scientists, allowing for more effective engagement. Advisory committees comprising members of relevant stakeholder groups, including vulnerable or underserved groups, could serve as representatives for their communities and could help identify strategies for more effective communication and engagement.

Barrier 5. There can be difficulties in establishing two-way information flow between scientists and stakeholders, due to one or both parties failing to see the value of communication. Moreover, there are challenges involved in coordinating diverse entities and individuals for any particular research effort, especially when there are numerous people and groups involved.

Opportunity 5. Role-playing exercises may help ensure that scientists, stakeholders, and others see the value of two-way communication. Efforts can be made to demonstrate the effectiveness and value of community engagement through case studies and storytelling, as a first step toward further engagement. Clear lines of communication, chain of command, and protocols, as well as the involvement of boundary spanners or boundary organizations, may facilitate the coordination of stakeholders and scientists in concerted efforts and will help participants feel involved and useful and have a sense of ownership.

Scientist Perspective

Barrier 6. Scientists' engagement with stakeholders can be limited by competing demands on time and by the relative importance placed on this engagement, in terms of promotion and professional recognition. In addition, scientists are often not trained in speaking to public audiences or engaging with stakeholders. They may also not be equipped to efficiently transfer knowledge or to provide appropriately tailored information to stakeholders.

Opportunity 6. Strong relationships, collaborations, and clear communication between scientists and stakeholders help produce scientific results that are most applicable to coastal decision making. To help facilitate the development of key relationships, funding programs could provide funds for engagement and knowledge-transfer activities and consider ways to incentivize collaborations between scientists and stakeholders via boundary organizations and other boundary spanners. Notably, there is an opportunity for extension faculty associated with sea- and land-grant programs to play a prominent role in future engagement and knowledge transfer between scientists and stakeholders. There may also be opportunities to offer training for scientists on effective communication and collaboration with stakeholders. Involving university leadership in these opportunities may also increase interest in and support for scientist engagement with stakeholders.

Barrier 7. Scientists working on or wanting to engage in research relevant to the Gulf Coast but who are not from or based there may feel limited by their "outsider" status when attempting to engage with stakeholders. They may have concerns about whether their information and expertise will be dismissed, especially if the information is viewed as contrary to deeply held stakeholder views.

Opportunity 7. Funding programs that focus on Gulf Coast-related research could encourage and facilitate collaborations among regional scientists (especially those with well-established relationships with stakeholders) and those from outside the region with complementary interests and expertise. To progressively build trust and forge strong collaborations, workshops, personnel exchanges, and symposia could be used to initiate communication and discussions among Gulf Coast stakeholders, Gulf Coast-based scientists, and scientists from outside the Gulf Coast who have relevant research interests and expertise.

CHAPTER 1

Introduction

The U.S. coastline is a dynamic and important region that hosts a large percentage of the population, has a critical role in the economy, and is made up of physically, geologically, ecologically, demographically, economically, and socially diverse environments (NOAA, 2013). With the number of people living in coastal areas increasing every year (NOAA, 2013), long-term planning to sustain healthy, productive U.S. coastlines and coastal communities depends on an understanding of the natural processes and human activities—and their interactions and feedbacks—that shape and change the coastal zone. Long-term changes resulting from these processes may affect the coasts' future habitability, health, and productivity.

The U.S. Gulf of Mexico Coast (hereafter referred to as "the Gulf Coast") provides a particularly relevant setting to study deeply connected natural and human interactions. This is in large part because of the historical, and continuing, concentration of a wide range of infrastructure and coastal development, much of it energy related, within a region that is highly vulnerable to hurricane landfalls and is experiencing the most rapid rise in relative sea level in the United States (Marcy et al., 2012, 2018). This confluence of intense, varied coastal development and rapid change in environmental stressors makes the Gulf Coast a harbinger of what is to come globally as the rates of sea level rise approach those at the end of the last glacial period—a time when many coastal systems were unable to keep pace and rapidly migrated landward.

Over various timescales, changes to the physical landscape, including natural and human-induced coastal subsidence and erosion; river, delta, and inlet migrations; and modifications from human activity, have significantly and in some places irreversibly changed the coastline. The construction of engineered infrastructure, such as levees and navigation channels, has affected delta morphology and the health and salinity of coastal marshes and related ecosystems. River systems that empty into the Gulf carry sediment and nutrient loads that have also changed over time due to agriculture, engineering, changes in the built environment, and other human activities. Such modifications can alter the natural (physical and ecological) system, often prompting people to make further modifications, thus creating a cycle that, over time, increases the likelihood that infrastructure, communities, and ecosystems will sustain damage due to sea level rise, coastal storms, and further development.

Over the next 10 to 200 years, the Gulf Coast shoreline will respond to a number of different environmental stresses, such as accelerating relative sea level rise, increasing frequency of intense hurricanes, and warming temperatures (e.g., Bender et al., 2010; NOS, 2011; Biasutti et al., 2012; Wright et al., 2015; Yan et al., 2017). Some of these stresses will lead to dramatic changes along the coast. For example, rapid sea level rise can lead to the loss of barrier islands, thus exposing the mainland to more energetic waves; the loss of wetlands due to their inability to keep up with rising seas; or abandonment of coastal communities and infrastructure when the cost of coastal protection becomes too great. These "tipping points" (discussed in more detail in Chapter 2) may be reached earlier along the Gulf Coast than in other U.S. coastal settings, due to its low-lying sedimentary coastline and vulnerable ecosystems.

Another stressor for the Gulf region is coastal storms. Widespread flooding and associated coastline erosion and ecosystem disturbance[1] during the 2017 Atlantic hurricane season led to major disruptions in energy, water supply, and wastewater treatment throughout the Gulf of Mexico, the Southeast Atlantic, and Caribbean coastal zones[2] (e.g., Egan, 2017; Jaimes et al., 2017; Jermoe, 2017; Halverson, 2018; Smith et al., 2018). Cascading impacts of the 2017 hurricane season—on commerce, the political landscape, and socioeconomic vulnerability, for example—are as yet unknown, but based on impacts from the past decade (e.g., the 2005 hurricane season), will likely be significant (Halverson, 2018). One consequence of the recent storm season was the outmigration of hundreds of thousands of people from Puerto Rico to the U.S. mainland (Alvarez, 2017); whether it is permanent or temporary remains unknown. Recent work by Hauer (2017) predicts that by 2100, sea level rise could lead to the outmigration of tens to hundreds of thousands in some Gulf Coast counties. Such an egress could tax inland urban centers such as Atlanta, Denver, and Phoenix (Hauer, 2017).

Currently, there are coastal ecosystem and shoreline restoration initiatives in place (e.g., the State of Louisiana's *2017 Coastal Master Plan*). In the case of Louisiana, however, these plans are only being implemented after significant and irreversible land loss has already occurred. Even if Louisiana's *Coastal Master Plan* is fully implemented, substantial net land loss will continue to occur over the next half-century (LACPRA, 2017). Thus, the Gulf

[1] For example, see pre- and post-Hurricane Harvey comparisons from Texas: https://www.usgs.gov/center-news/pre-and-post-storm-photo-comparisons-texas?qt-news_science_products=2#qt-news_science_products.

[2] *Response and Recovery to Environmental Concerns from the 2017 Hurricane Season: Hearing Before the Members, Subcommittee on the Environment,* 115th Congress. 2017. Statement of the Majority Staff of the Committee on Energy and Commerce.

Coast offers a unique opportunity to gain fundamental understanding of coastal system evolution—both human and natural—that will be globally relevant. A major scientific challenge, with large societal implications, is whether projections of the conditions under which such coastal tipping points are reached can be improved.

Solutions to the Gulf Coast region's contemporary problems can be informed by improved understanding of the potential evolution of the system. Though still challenged by uncertainty, such understanding could enable objective evaluation of long-term outcomes of different mitigation approaches and thus allow assessment of different courses of action. The recent past and future evolution can only be understood, projected, and affected if the physical, ecological, and human components of the system, as well as their interactions and feedbacks, are well understood.

STUDY ORIGIN

The National Academies of Sciences, Engineering, and Medicine's (the National Academies') Gulf Research Program (GRP) asked the National Academies' Board on Earth Sciences and Resources, Ocean Studies Board, and Board on Environmental Change and Society to undertake a study on long-term coastal zone dynamics and the interactions and feedbacks between human and natural processes that occur in the coastal zone. The complete Statement of Task (SOT) is provided in Box 1.1. The project originated as a way to better understand the multiple factors that influence long-term processes along the Gulf Coast, and was envisioned to help inform decision making and planning related to the four strategic initiatives of the GRP:

- *Healthy Ecosystems* – Advance understanding of ecosystem processes and dynamics to facilitate sustainable use of natural resources.
- *Thriving Communities* – Enable people and coastal communities to successfully prepare for, respond, and adapt to stressors and adverse events.
- *Safer Offshore Energy Systems* – Foster minimization and management of risk to make offshore operations safer for both people and the environment.
- *Capacity Building* – Enhance the ability of researchers, decision makers, and communities to use science to solve challenges at the intersections of human, environmental, and offshore energy systems.

The committee believes that, in addition to informing the GRP, this report can assist with possible research directions and decision making among relevant scientific disciplines, private-sector organizations, and local, state, and regional governments.

BOX 1.1
Statement of Task

An ad hoc committee will conduct a study to determine research needed to improve understanding of the interactions and feedbacks between long-term (decadal to millennial scale) natural coastal dynamics and development, including energy-related infrastructure, along the U.S. Gulf of Mexico coastline. The work will be informed by case studies from other U.S. regions, including the mid-Atlantic, California, and/or Alaska.

Recognizing the Gulf Research Program's interest in understanding the Gulf of Mexico region's interconnected human, environmental, and energy production and development systems, the study will:

1. Identify gaps in scientific and technical understanding of the interactions and feedbacks between physical processes and coastal development in the U.S. Gulf of Mexico, and similarities and differences in these processes compared to other U.S. coastlines.
2. (a) Define the essential components of a research and development program (e.g., monitoring, data collection and management, modeling, population surveys, multi-dimensional mapping) in response to the gaps identified in (1); and
   (b) Develop and set priorities for no more than three critical areas of research to increase understanding of long-term natural coastal dynamics (e.g., sea-level rise; coastal subsidence, uplift, and erosion; coastal ecosystem evolution; coastal hazards) in order to advance the science and help inform stakeholder decision making, especially for those activities focused on energy and related infrastructure.
3. Identify barriers to, and opportunities for, more effective communication among scientists and coastal stakeholders about improved monitoring, forecasting, mapping, and other data collection and research regarding long-term changes in U.S. coastlines.

STUDY SCOPE AND REPORT APPROACH

The Committee on Long-Term Coastal Zone Dynamics: Interactions and Feedbacks Between Natural and Human Processes Along the U.S. Gulf Coast was convened to answer this request. Twelve committee members brought to this task a broad spectrum of knowledge and expertise related to coastal geology, ecology, and engineering; geography; energy systems development; economics; and social and behavioral science. Committee member and staff biographies are provided in Appendix A.

Toward the beginning of the study process, the committee determined that several terms in or related to the SOT needed a common definition to better define the report scope. Definitions for these key terms are found in Box 1.2. Because of the emphasis on energy and energy-related infrastructure in the SOT, the committee's approach was to focus the report content slightly more toward the Louisiana and Texas coasts, which have a higher

BOX 1.2
Definitions

The **coastal zone** is generally defined by its landward and seaward boundaries. For this report, the committee adopted a seaward boundary based on the influence of storm waves on measurable changes to the sea bottom, namely a boundary inshore of which wave motion drives sediment transport. This depth is typically between 15 and 40 meters, which is also the depth range for initiation of significant storm surge generation. Definition of the landward boundary is fluid because its relevance depends on the specific scientific questions and policy issues being addressed. For modeling some aspects of the physical system, the landward boundary of the coastal zone would be the farthest inland extent of storm surge (surge + tide + wave setup + wave swash) on the mainland. As illustrated by Hurricane Harvey, planners need a landward definition that captures the impact of precipitation-related flooding interacting with surge. Moreover, political boundaries are important because they influence the tax base for adapting to sea level rise and storms.

Long term is defined as 10 to 200 years in the context of this study. This timeframe reflects the extent to which sea level rise projections have a reasonable degree of confidence, while reflecting the anticipated design lifespan of coastal engineering structures, energy-related infrastructure, and residential and commercial development. This includes consideration of how episodic events may affect long-term changes driven by both natural (e.g., storm intensity and frequency, ecosystem composition changes, sea level rise) and human processes (e.g., human decision making, demography, coastal development, changes in energy needs and sources), as well as the cumulative effects of short-term events superimposed on longer-term forcing. The committee uses two timescales in the report: near-decadal (10–50 years) and decadal to century (50–200 years).

For this report, **coastal development** includes infrastructure that directly or indirectly supports the economy (including the energy sector) along the Gulf Coast, including residential and commercial buildings, schools, hospitals, and roads.

Energy infrastructure is the physical infrastructure required for producing, transforming, transmitting, distributing, and storing energy associated with oil and natural gas (based on the definition used in Goldthau, 2014). This definition does not exclude, but puts secondary focus on, infrastructure associated with other energy systems, including nonrenewable sources such as coal and nuclear, and renewables (e.g., wind, waves, solar, tidal). Because the report is focused on the nearshore coastal environment, the committee did not consider offshore oil and gas infrastructure, but included infrastructure that lies within a state's jurisdiction and spans the coastal zone (e.g., offshore oil and gas pipelines that terminate landward of the Outer Continental Shelf boundary, Gulf Intracoastal Waterway). Fixed infrastructure systems such as roads, ports, and railroad tracks were also included.

prevalence of such infrastructure compared with Alabama, the Gulf Coast of Florida, and Mississippi. However, there are discussions of and examples from all of the Gulf states throughout the report and most of the processes addressed are relevant to all Gulf states.

During the study, the committee convened three information-gathering meetings (May 17–18, 2017, Washington, DC; September 18–19, 2017, New Orleans, Louisiana; November 15–16, 2017, St. Petersburg, Florida); one workshop with 20 invited participants (July 18–20, 2017, Houston, Texas); and an additional meeting in closed session to develop this report (January 18, 2018, Galveston, Texas). The committee heard presentations from leaders in fields related to many aspects of the SOT, including state and federal agencies, academia, nongovernmental organizations, and the energy industry. These participants are listed in Appendix B. The committee also consulted peer-reviewed research literature, community-sponsored efforts, and state and federal government reports to provide a strong scientific foundation.

The report structure is as follows: Chapter 2 summarizes the current state of the scientific knowledge about the Gulf Coast and its processes; Chapter 3 discusses the high-priority gaps in scientific knowledge associated with these processes (hereafter referred to as "research gaps") (SOT Task 1); Chapter 4 presents barriers to communication between scientists and stakeholders (SOT Task 3); and Chapter 5 identifies critical areas of research to increase understanding of long-term coastal dynamics (SOT Task 2b), presents guidelines for a research agenda (including essential components of a research and monitoring program [SOT Task 2a]), and opportunities for more effective communication between scientists and stakeholders (SOT Task 3).

The committee hopes this study will provide valuable guidance for the GRP and an array of governmental and nongovernmental stakeholders.

CHAPTER 2

Background of the Gulf Coast System

The Gulf of Mexico is a marginal sea of the Atlantic Ocean, with substantial mixing of Atlantic and Gulf waters (see Khade et al., 2017, for a recent discussion), including inflows of saltwater from the Caribbean Sea and freshwater from two-thirds of the conterminous United States and a portion of Canada. Formed 165 to 142 million years ago as seafloor spreading continued following the breakup of the supercontinent Pangea (Stern and Dickinson, 2010), the Gulf had opened to the Atlantic Ocean approximately 155 million years ago. Over millions of years, the deposition, subsidence, and chemical alteration of organic-rich sediments within the historical Gulf of Mexico basin resulted in the creation and storage of vast energy resources. Through more than a century of exploration, major hydrocarbon (oil and natural gas) quantities have been found in much of the basin's sedimentary layers, particularly in areas now located in the northern[1] and northwestern[2] Gulf of Mexico (Galloway, 2008).

The Gulf coastline consists of a range of different types of coastal settings (see Figure 2.1), including barrier islands and peninsulas, tidal inlets and tidal deltas, rivers and river deltas, cheniers, bays, marshes, cypress swamps, and mangrove forests, all of which respond differently to changing environmental and anthropogenic stressors. About two-thirds of the northern Gulf Coast consists of barrier islands or spits (e.g., Morton et al., 2004), low-lying, sandy landforms backed by coastal bays or lagoons and separated from each other or the mainland shore by tidal inlets. Because barrier islands are primarily composed of shifting sand, the structure and composition of each island is unique (Pilkey, 2003). The remainder of the coastline consists of mainland shore without an intervening bay or lagoon, including sandy beaches and bluffs and marsh shorelines. The Gulf Coast

[1] The northern U.S. Gulf of Mexico extends from Galveston Bay, Texas, to the mouth of the Suwannee River, Florida.

[2] The western U.S. Gulf of Mexico is the Texas coastline, west of approximately 94° longitude. The central Gulf coast encompasses Louisiana, approximately 94° to 89° longitude. The eastern Gulf refers to Mississippi, Alabama, and Florida, approximately 89° to 83° longitude.

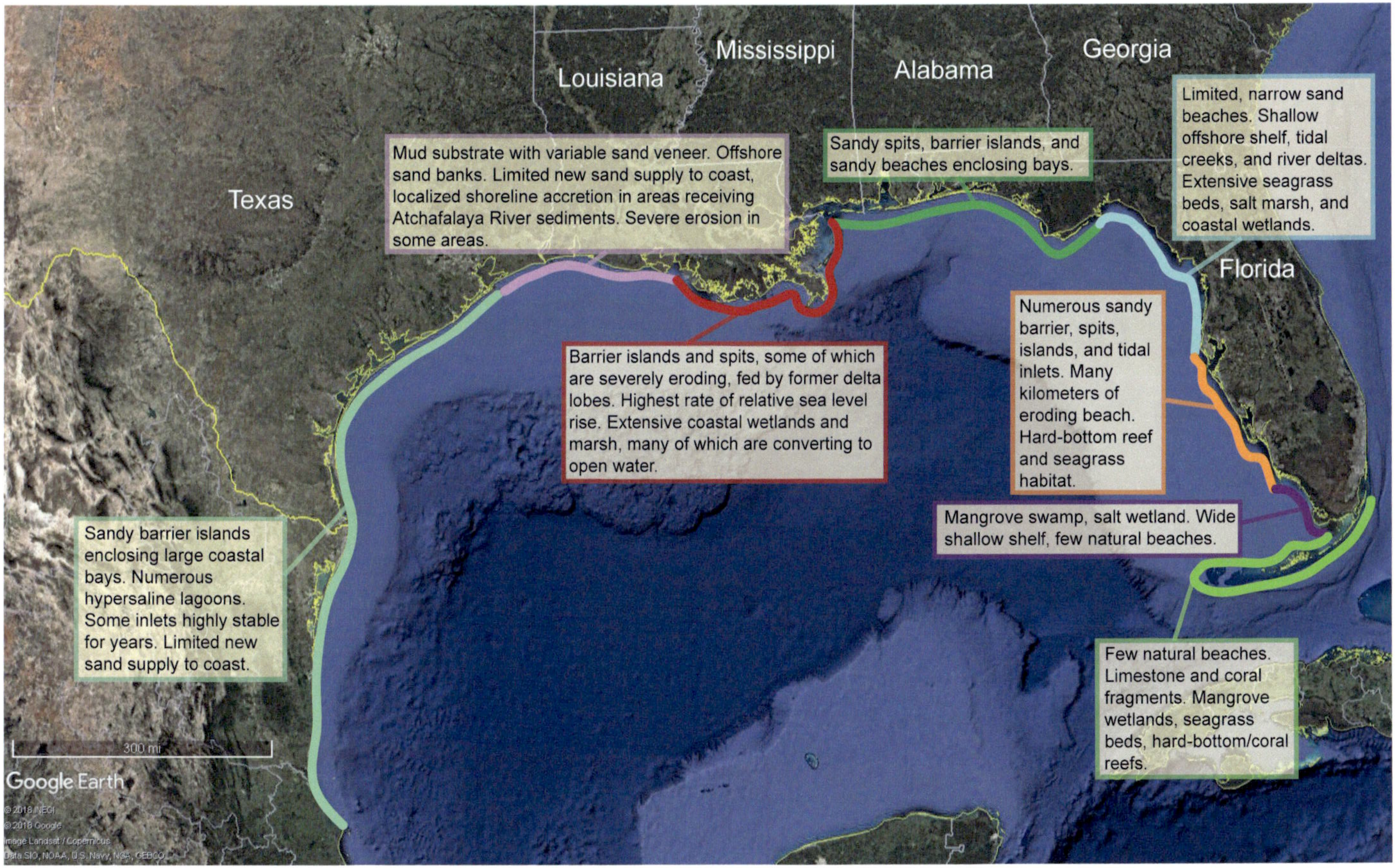

FIGURE 2.1 General characteristics of Gulf Coast geomorphology, geology, and ecology. SOURCE: Modified from Pendleton et al., 2010.

includes many large estuarine systems (e.g., Barataria Bay, Corpus Christi Bay, Galveston Bay, Mobile Bay, Pensacola Bay, Sabine Lake, Tampa Bay) and several hundred small estuaries (EPA, 1999). Many of the large bays (e.g., Mobile Bay, Alabama) are drowned river valleys that flooded as sea level rose following the Last Glacial Maximum or are associated with inter-distributary bays in the Mississippi River Delta.

The Gulf barrier islands, along with other coastal landforms such as most modern major river deltas, estuaries, and coastal wetlands, were formed approximately 5,000 to 6,000 years ago as sea level rise slowed during the late Holocene Epoch (Stutz and Pilkey, 2011; Anderson et al., 2014). Most evolved along non-deltaic shorelines, while a smaller number (e.g., the Chandeleur Islands, Louisiana) are younger, having formed along deltas dominated by tidal and wave energies and characterized by moderate supplies of fluvial sediments (Penland et al., 1985, 1988; Otvos and Carter, 2013). Given some debate among investigators regarding the characteristics that distinguish barrier islands from other types of coastal islands (Otvos, 2012), the Gulf of Mexico includes 116 barrier islands of both non-deltaic (coastal plain) and deltaic origins, representing approximately 5 percent of the world's total number and a combined island length of 2,398 kilometers (Stutz and Pilkey, 2011).

People have lived along the Gulf Coast for at least 14,500 years (Halligan et al., 2016). It is not known whether human modifications to the physical landscape during the pre-colonial period altered coastal geomorphology in significant ways, but there is clear evidence that people in the colonial period altered the local geomorphology in the coastal zone. In early 18th-century New Orleans, the French built a mile-long bulwark on top of a natural levee. By 1763, 50 miles of waterfront had levees (Colten, 2005). Large reclamation projects, shoreline armoring, canal building, and port and channel dredging did not begin along the western Gulf Coast until the 19th and 20th centuries. These much more extensive changes broadly transformed the landscape and altered physical and ecological processes.

Today, the Gulf Coast states—Alabama, Florida, Louisiana, Mississippi, and Texas—are home to approximately 19 percent of the U.S. population.[3] The Gulf Coast is generally less urbanized than the rest of the coastal United States, and coastal counties are less densely populated compared with the U.S. average coastal county population density (Wilson and Fischetti, 2010). However, major metropolitan regions exist near Houston and Galveston, Texas; Tampa Bay, Florida; and to a lesser extent, around New Orleans, Louisiana. Since 1960, the population in most Florida Gulf Coast counties (including the Tampa Bay area), as well as the Houston metropolitan region, has increased at a rate faster than the U.S. population (Wilson and Fischetti, 2010; NOAA, 2013; U.S. Census Bureau, 2018). With

[3] Percentage calculated from 2017 population estimates available at: https://www.census.gov/data/tables/2017/demo/popest/state-total.html.

a few exceptions, the remainder of the Gulf Coast counties increased at a rate slower than the U.S. average, or even decreased in population. The demographic and socioeconomic profile of Gulf Coast residents largely mirrors the U.S. population as a whole, but with somewhat greater ethnic diversity and larger income disparities than other regions. The Gulf Coast population tends to be older, earns a lower median household income, and includes a greater share of African American residents compared with the U.S. national average (NOAA, 2011). Gulf Coast communities also show higher poverty rates than the nation as a whole (NOAA, 2011). A growing literature suggests that these characteristics are associated with greater social vulnerability to the effects of sea level rise, hurricanes, and other hazards—meaning that they are less able to prepare, respond, rebuild, or relocate before or after disasters (Cutter and Emrich, 2006; Cutter et al., 2006, 2008; Picou and Marshall, 2007).

An important driver of human interaction with the environment, and a significant reason behind population growth and many of the modifications to the landscape and coastline over the past century, involved infrastructure related to the extraction of energy resources. President Truman's declaration of federal jurisdiction over the continental shelf led to the granting of offshore oil and gas leases, beginning in 1947. In 1953, the Submerged Lands Act granted states jurisdiction from the high-tide line to 3 miles offshore. These policies led to a significant increase in the construction of oil and gas infrastructure within the coastal zone (especially along the northern and western Gulf states), including production and processing facilities, canals dredged for providing access to construction equipment, land reclamation by draining and filling of wetlands, and the development of supporting transportation infrastructure such as deep water navigation channels and ports. The latter expanded dramatically in the post–World War II period, with the biggest changes in the 1950s to late 1970s (Theriot, 2014). As a result, the Gulf Coast has historically been, and continues to be, a dominant contributor to the nation's crude oil and natural gas production. More than 51 percent of total U.S. petroleum refining capacity, and more than 50 percent of total U.S. natural gas processing plant capacity, is located along the Gulf Coast (EIA, 2017b,c). There is a vast amount of infrastructure associated with not just production and processing facilities, but through the entire oil and gas supply chain, including storage (e.g., aboveground and underground tanks) and transportation and distribution (e.g., pipelines, ports, canals, waterways, roads, railroads) (see Figure 2.2; CCSP, 2008; Dismukes, 2011; Needham et al., 2012; Theriot, 2014). The result is an extensive and complex infrastructure that developed to meet national and global demands for the exploration, extraction, transport, and use of Gulf of Mexico energy resources.

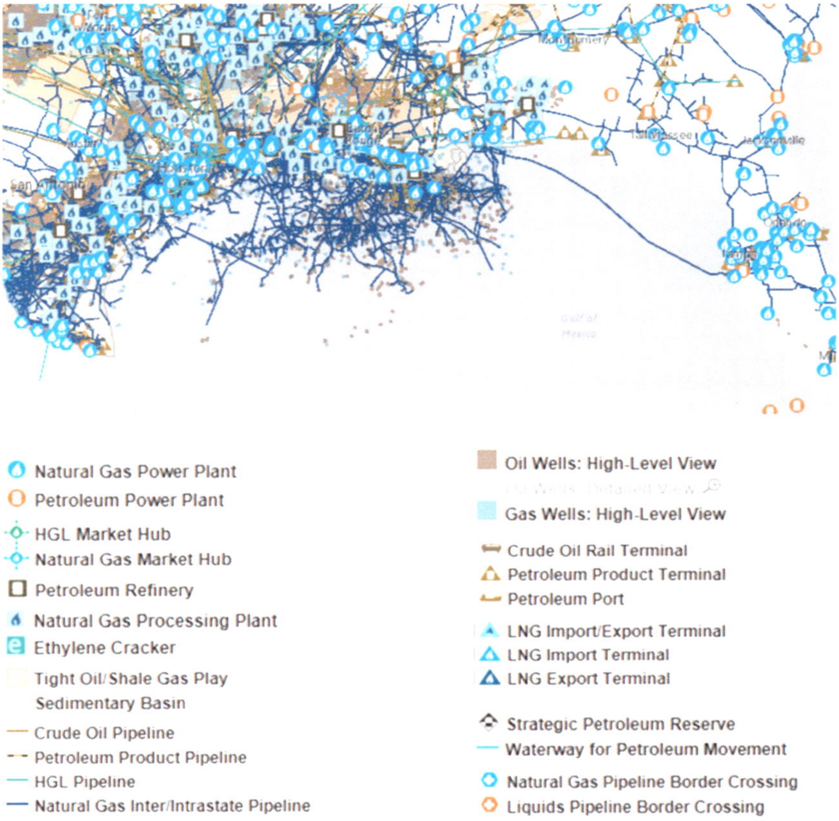

FIGURE 2.2 Major components of oil and gas infrastructure along the Gulf Coast. SOURCE: EIA, 2017a.
NOTE: HGL refers to hydrocarbon gas liquids; LNG refers to liquefied natural gas.

THE COUPLED NATURAL-HUMAN SYSTEM

A common theme in the Gulf Coast's environmental history is the importance of interactions and feedbacks between and within the natural (physical and ecological) and human systems. An interaction exists when system *A* affects system *B* or system *B* affects system *A*, while a feedback exists when system *A* affects system *B*, and system *B*, in turn, affects system *A* as a two-way interaction (see Box 2.1).

Examples of physical-ecological *interactions* in the coastal zone are numerous and include loss or gain of coastal habitats in response to sea level rise (e.g., Day et al., 2008; Kirwan and Megonigal, 2013), coastal storm inundation of wetlands (e,g., Howes et al., 2010; Tweel and Turner, 2012a), contribution to wetlands by sediment flow from a large

BOX 2.1
Interactions and Feedbacks

This box draws on a wide range of sources from the coupled human-natural system literature (e.g., Turner et al., 2003; Kennedy, 2006; Liu et al., 2007; Werner and McNamara, 2007; Smith, 2008; Carpenter et al., 2009; Horan et al., 2011; Brashares et al., 2014; Smith, 2014; Withey et al., 2014). Suppose there are two systems, *A* and *B*. An **interaction** exists when system *A* affects system *B* or system *B* affects system *A* (Figure a). A **feedback** exists when system *A* affects system *B*, and system *B*, in turn, affects system *A* (Figure b). When there is a feedback, there is always an interaction. However, when there is an interaction, there is not necessarily a feedback.

Feedbacks can be **positive** or **negative.** Positive feedbacks are self-reinforcing such that an increase in *A* triggers an increase in *B*, and the increase in *B* triggers a further increase in *A*. Negative feedbacks involve offsetting or equilibrating effects, potentially leading to establishment of an equilibrium state. An increase in *A* triggers an increase in *B*, but the increase in *B* triggers a decrease in *A*. The resulting decrease in *A* triggers a decrease in *B* and a subsequent increase in *A*.

A fishery illustrates the concept of a coupled natural-human system. The natural system is the stock of fish and the human system is the number of fishing vessels (Figure c). A larger fish stock makes fishing more profitable, more fishers enter the fishery, and so there are more fishing vessels. With more vessels, fishers deploy more fishing gear and exert greater pressure on the resource and reduce the fish stock. As the fish stock declines, fishing becomes less profitable, vessels exit the fishery, and the fish stock rebounds. Because of the profit incentive, fisheries are generally characterized by negative feedbacks (Smith, 2014). Ocean warming is expected to influence the distribution of fish stocks globally (Burrows et al., 2011). As such, warming is an important interaction that influences the natural system (Figure d). Although fishing vessels burn fossil fuels and thus contribute to greenhouse gas emissions, which, in turn, influence climate change and associated ocean warming, for any given fishing fleet, the contribution of greenhouse emissions to ocean warming is trivially small.

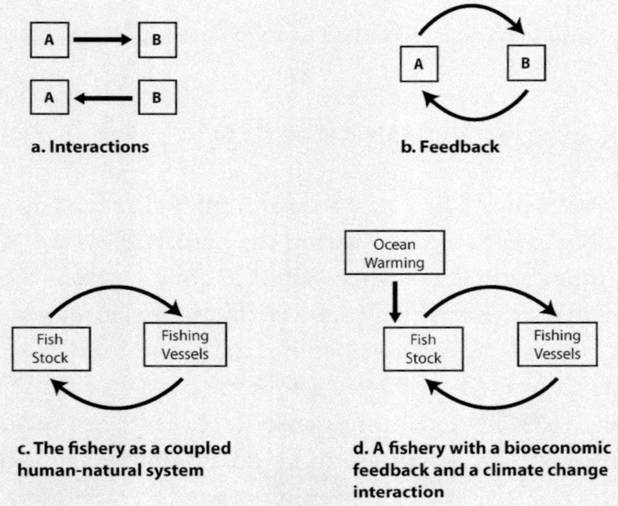

river (Paola et al., 2011; Kolker et al., 2012; Esposito et al., 2013), damping of waves by vegetation (e.g., Kobayashi et al., 1993) and corals (Lugo-Fernández et al., 1998), reduction of storm surge by vegetation (Wamsley et al., 2010), and loss of barrier island ecosystems in response to sea level rise and low shoreface sand content (e.g., Fearnley et al., 2009; Otvos and Carter, 2013; Moore et al., 2014). Irish et al. (2010) showed increased back-bay flooding during coastal storms following barrier island dune degradation. There are also various physical-ecological *feedbacks*. An example is the alteration of barrier island topography as sea level rise and increasingly episodic inundation lead to loss of dune habitats and lower island topography, which is then more vulnerable to inundation (i.e., a self-reinforcing feedback; e.g., Durán and Moore, 2015).

The feedbacks between the natural and human components of the coastal system are especially important for understanding future coastal evolution. As natural coastal changes occur, people respond by making modifications, which in turn alter evolution of the natural system, thereby further influencing human behavior. A classic example of such a feedback is often set in motion by coastal erosion. In response to erosion, protective infrastructure (e.g., a seawall) is installed, which reduces erosion rates, leading to more coastal development because the risk of storm damage has been reduced. The seawall could also affect sediment dynamics at neighboring communities, possibly causing more erosion and triggering further installation of infrastructure. These types of feedbacks between the human and natural systems are discussed in more detail in the last section of this chapter.

Tipping Points

An important concept used in modeling coupled systems is that of a *tipping point*, a critical point "at which a sudden and dramatic shift to a contrasting dynamical regime may occur" (Scheffer et al., 2009, p. 1). The long-term dynamics of the Gulf Coast raise the possibility of tipping points in the coupled ecological-physical system and provide a way to consider the coupled natural-human system. For example, continued sea level rise and changes in sediment supplies could eventually trigger a breach in a barrier island. While the initial breach area would be relatively small, given limited sediment supply and rising sea level the breach would very likely continue to widen, allowing more water in and out of the system, eventually leading to substantial land loss. In this way, a relatively small event (e.g., a breach) could lead to substantial changes in hydrodynamics that would cause a significant qualitative change in the coupled ecological-physical system because it would radically alter salinity and expose coastal marshes and other mainland habitats to waves, tides, and storm surge (Culver et al., 2007).

Large breaches or a total loss of barrier islands could also lead to tipping points in the coupled natural-human system; for example, the complete disappearance or relocation of the community occupying the island, or major shifts in flood frequency for a community behind the barrier island. In contrast to incremental outmigration in response to sea level rise (Hauer, 2017), a tipping point in this context is a point of no return, a critical juncture after which a coastal community eventually will cease to exist in its original location.

Even without full-scale disappearance of barrier islands, physical changes in the coastal zone could lead to tipping points in the coupled natural-human system. For example, as sea level rise and wetland loss continue, many oil and gas pipelines continue to be maintained even though their vulnerability to wave action increases. However, a tipping point may be reached after which further incremental change in sea level or wetland loss leads to increased damage and associated maintenance costs, triggering abandonment of some of these assets with resulting implications for cleanup costs and the need to move or store energy resources differently. Another tipping point relevant to the energy sector is the potential for a qualitative change in the energy workforce along the Gulf Coast. As sea level rise contributes to outmigration and land loss increases areas of open water, the labor force who can commute by car to oil and gas processing and distribution facilities in the coastal zone shrinks. Eventually, the labor force may rely on ships and aircraft to reach these facilities, increasing costs for facility operators who could, depending on the price of oil, cause their abandonment.

Decision Making Under Deep Uncertainty

Making very detailed or accurate predictions about exactly how the Gulf Coast will evolve over the next 10 to 200 years is not, of course, a realistic possibility. The uncertainties (both human and natural) over that period are simply too large, and outcomes will depend in large part on policy decisions occurring at the national or the international level (e.g., carbon emissions mitigation, global economic trends). It may be difficult or even impossible to characterize these long-term uncertainties using probability distributions, or there might be substantial disagreement regarding the likelihood of different outcomes. This is sometimes referred to as *deep uncertainty*, and it challenges decisions made in the near term that will nevertheless shape the long-term future (Lempert et al., 2003).

Improving understanding of the Gulf Coast coupled system and its interrelationships, however, can support better near-term decision making, even when faced with deep uncertainty. Researchers and practitioners can use an "exploratory" approach when developing and applying simulation models to conduct future scenario analysis (Bankes, 1993). This approach uses quantitative models as platforms to ask "what if?" questions rather than seeking to predict the future, providing a way to plausibly link different assumptions about future uncertainties to outcomes across the Gulf Coast coupled system, as well as to consider how policy changes or new investments might perform against a wide range of future conditions.

The field of decision making under deep uncertainty (DMDU) has evolved from this basic premise. DMDU methods include systematically testing policy choices or infrastructure plans against a wide range (i.e., thousands to millions) of plausible futures and identifying those plans that are more robust, meaning they perform well regardless of how the future plays out (Groves and Lempert, 2007; Hallegatte et al., 2012; Walker et al., 2013; Herman et al., 2015). Often, this entails creating adaptive plans that specify adjustments over time in response to new information, as well as key tipping points or conditions that could be monitored to inform these adaptive pathways (Kwadijk et al., 2010; Groves et al., 2013; Haasnoot et al., 2013). DMDU approaches, such as Robust Decision Making, also help identify "low-regret" choices that are insensitive to future uncertainty, as well as "decision-relevant" scenarios that illuminate key tradeoffs and can help to inform policy deliberations (Bryant and Lempert, 2010).

THE NATURAL SYSTEM

Physical Processes

The Gulf Coast is a complex region, governed by a combination of physical drivers originating in the ocean, in the atmosphere, and on land. The Gulf Coast itself is a microtidal environment, with astronomical tides of 1 meter or fewer in most places (NOAA, 2018). The region is dominated by broad, shallow continental shelves—except near the mouth of the Mississippi River, which discharges into deep water. Given the region's mostly shallow bathymetry, estuaries and many nearshore coastal environments are often well mixed. However, intense stratification occurs around the Mississippi River plume, where a lens of freshwater several meters thick can develop (Walker, 1996; Kolker et al., 2014). The Loop Current, a geostrophic current that originates in the southern Gulf and can extend northward, is a dominant feature of the pelagic zone and can transport heat and moisture northward to the coast (NASEM, 2018). Climatologically, the region is dominantly subtropical to tropical, with warm temperatures that fuel atmospheric convection and intense thunderstorms (particularly in the summer) (Kumpf et al., 1999). During summer and autumn, the entire Gulf Coast is prone to experiencing tropical cyclones, which can be the most severe driver of physical processes in the region (Trenberth et al., 2007; Miner et al., 2009). The northern Gulf is also prone to cold fronts, which strike on an approximately weekly basis between October and April (Roberts et al., 1989; Allison et al., 2000). These can be important drivers of mixing, sea level change, and coastal erosion.

Sea Level Rise

Low-elevation coastal zones such as the Gulf Coast face increased vulnerability, now and in the future, as a result of accelerating rates of relative sea level (RSL) rise (Nerem et al., 2018). RSL rise is determined by global sea level rise plus change in land elevation associated with subsidence. These processes give rise to spatially and temporally variable patterns of local RSL change.

During the Last Glacial Maximum (21,000–26,000 years ago), global sea level was about 135 meters lower than the level observed today (Lambeck et al., 2014) because ocean water was locked up in continental ice sheets, mainly those that covered large parts of North America and Eurasia. The Antarctic and Greenland ice sheets, as well as mountain glaciers, were also considerably larger during the Last Glacial Maximum. The transition into the present interglacial period and associated global sea level rise occurred between 21,000 and 7,000 years ago—its end coincided with the demise of the Laurentide Ice Sheet in North America. Despite the slowdown in the rate of global sea level rise, which persisted until the early 20th century (Gehrels and Woodworth, 2013), the Gulf Coast has seen continuous RSL rise due to subsidence associated with glacial isostatic adjustment. Figure 2.3 illustrates RSL curves for selected Gulf Coast regions, obtained from geophysical modeling of glacial isostatic adjustment and validated by geological RSL data.

Modern average, long-term RSL rise rates determined from National Oceanic and Atmospheric Administration (NOAA) tide gauges along the Gulf Coast (including both changes in water level and changes in vertical land movement) range from 1.38 millimeters per year (mm/yr) at Apalachicola, Florida, to a maximum of 9.65 mm/yr at Eugene Island, Louisiana (Pendleton et al., 2010). These rates, however, may be higher or lower for periods of time, such as during times of high fluid extraction (Kolker et al., 2011; Jones et al., 2016). Satellite altimetry data from the Gulf of Mexico show a mean present-day sea level rise (i.e., rise in water level alone) rate of 2.0 ± 0.4 mm/yr (Letetrel et al., 2015), although spatial variability within this region is very high due to processes such as the migration of the Loop Current, which can be an important control on sea surface heights (Liu et al., 2016). Other studies (e.g., Rietbroek et al., 2016) have suggested that rates in the western Gulf of Mexico are higher. Since changes in RSL are affected by the rate of change in all of the processes mentioned above, each is addressed in turn in the following section.

Sea level is rarely stable because changes in ocean water mass and volume are ongoing. Changes in the cryosphere (glaciers and ice sheets) will have a profound impact on low-elevation coastal zones worldwide. Ice loss in Antarctica is of particular importance to the Gulf of Mexico, given that the largest sea level rise due to the melting of the West Antarctic Ice Sheet (the most vulnerable portion of the Antarctic Ice Sheet) is centered on the northwestern Atlantic (Mitrovica et al., 2009).

Global sea level has been rising since the mid-1800s, and significant global sea level rise has been recorded over the past 100 years (see Figure 2.4), to a notable extent due to thermal expansion and melting mountain glaciers. Global sea level projections from the most recent Intergovernmental Panel on Climate Change report (2013) are now considerably higher, to a large extent due to the recognition of ice sheet instabilities (DeConto and Pollard, 2016).

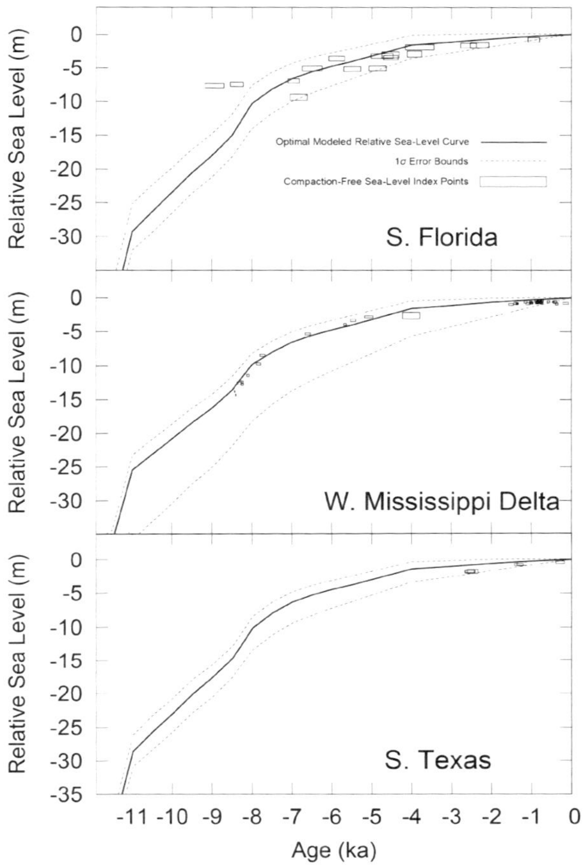

FIGURE 2.3 Relative sea level rise in the Gulf of Mexico over the past 12,000 years, as illustrated by discrete data (represented by error boxes) and geophysical model calculations (curves). SOURCE: Modified from Love et al., 2016.

Future sea level rise predictions suggest a rise in global sea level of 0.3 to 2.5 meters by the year 2100, with an intermediate-high value of 1.5 meters (Sweet et al., 2017).

Dynamical sea level adjustments also occur. These are changes in sea level that are driven by long-term (i.e., annual to multi-decadal) changes in winds, atmospheric pressure, ocean currents, and similar phenomena that shift the distribution of water masses (Kolker and Hameed, 2007; Sturges and Douglas, 2011; Liu et al., 2016). These shifts can be driven by year-to-year fluctuations in weather, internal modes of climate variability (e.g., the El Niño Southern Oscillation, the North Atlantic Oscillation, and the Atlantic Multi-Decadal Oscillation), and long-term changes in climate. Dynamical sea level can

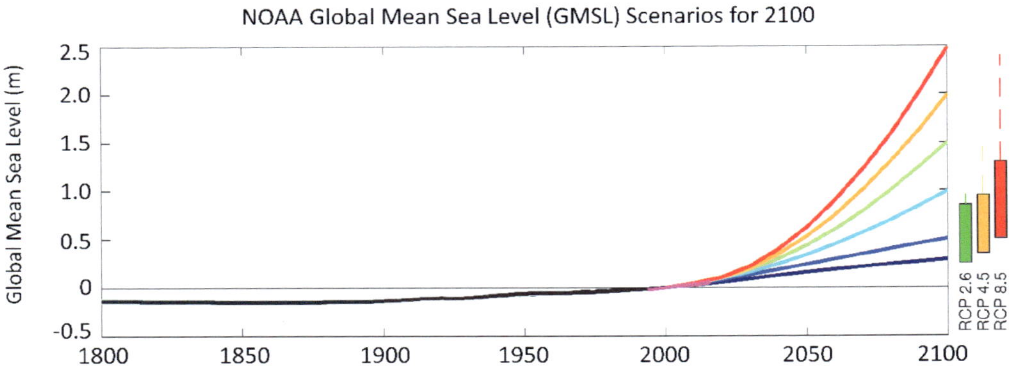

FIGURE 2.4 Six representative global sea level rise scenarios for the 21st century (the six colored lines represent low, intermediate-low, intermediate, intermediate-high, high, and extreme scenarios, respectively) relative to historical geological tide gauge and satellite altimeter reconstructions from 1800 to 2015 (black and magenta lines), along with central 90 percent conditional probability ranges (colored boxes) of representative concentration pathway-based (RCP) global sea level rise projections (Sweet et al., 2017, and references therein). The dashed lines above the central 90 percent probability ranges reflect the upward revision of projections due to improved understanding of the Antarctic contribution (DeConto and Pollard, 2016). SOURCE: Sweet et al., 2017.

contribute to variability and trends on multiple timescales. On a year-to-year basis, it can result in about +/– 10 centimeters of sea level change, which can potentially obscure tide gauge records that provide insight into climatically and subsidence-driven sea level rise (Kolker et al., 2011). Dynamical sea level is likely to contribute about 5 centimeters of total sea level rise for the Gulf Coast during the remainder of this century (Hall et al., 2016, and references therein).

Dynamical processes drive sea level change on shorter timescales as well. Loop Current migration is one driver and can be addressed with data-assimilative efforts (e.g., Hoteit et al., 2013). Another particularly important process is the cold front cycle, which occurs on approximately weekly timescales from October through April and leads to local temporary sea level rise on the order of 0.25 to 1 meters and temporary sea level drop on the order of 0 to 0.5 meters (Allison et al., 2000). The asymmetry in the size of the surge versus recession is often a function of the shallow nature of the system and the duration of the pre- and post-phases of a front (Roberts et al., 1989, 2015; Draut et al., 2005). While the magnitude of frontally driven sea level change is substantially less than a storm surge (e.g., up to ~10 meters during Hurricane Katrina), the regular occurrence of these events makes them among the most important current short-term drivers of sea level change across the northern Gulf Coast.

Subsidence

Much of the Gulf Coast is subsiding, but the driving processes and their rates vary by several orders of magnitude, depending on the geographic location. Important drivers include tectonic processes, faulting, sediment loading, glacial isostatic adjustment, compaction, and fluid withdrawal (see Figure 2.5).

The Florida peninsula is tectonically stable, as it consists predominantly of Cenozoic carbonate bedrock. By comparison, the remainder of the Gulf Coast is underlain by thick clastic layers, dominantly mud and sand, that exhibit deep-seated subsidence rates of typically less than 0.25 mm/yr (e.g., Woodbury et al., 1973; Paine, 1993). Shore-parallel growth fault systems are very common along the Gulf Coast west of Florida (Murray, 1961). Few studies, however, have quantified fault slip rates. Recent studies have focused mainly on the Baton Rouge Fault Zone in Louisiana, with reported time-averaged slip rates of 0.2 to 1.2 mm/yr over the past ~4,000 years, although rates averaged over longer timescales (~30,000 to 130,000 years ago) are an order of magnitude lower (Yeager et al., 2012; Shen et al., 2017). An example of relatively recent fault activity, with maximum throw estimated

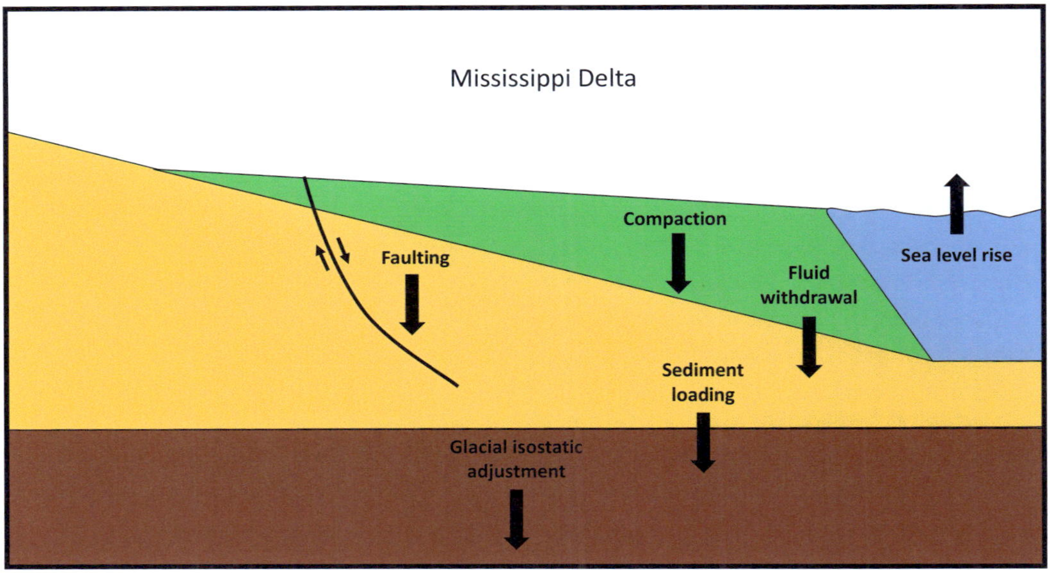

FIGURE 2.5 Driving processes for subsidence along the Gulf Coast.

at 75 centimeters over 40 to 50 years, was reported from a coastal wetland setting in central Texas (Feagin et al., 2013). The understanding of recent faulting in the region is an area of emerging research.

Glacial isostatic adjustment is related to the existence of the up to 4 kilometer thick Laurentide and Cordilleran ice sheets during the Last Glacial Maximum. The weight associated with these ice sheets led to a depression of the Earth's crust at their locations and an associated uplift in a large region surrounding the ice sheets. Once the ice sheets retreated, the areas that experienced uplift, including the Gulf Coast region, began to subside. Associated subsidence due to glacial isostatic adjustment along the Gulf Coast continues today at rates between 0.3 and 1.5 mm/yr (Love et al., 2016). Widely spaced tide gauge-derived rates do not account for the full range of variability in subsidence in the region; rates of RSL rise are likely higher in places with high rates of shallow subsidence such as in coastal Louisiana (Nienhuis et al., 2017). Compaction is an important process in settings that experience rapid sedimentation. It includes primary consolidation (i.e., the rapid release of porewater from the shallowest water-lain strata during burial with younger sediments) and secondary consolidation (i.e., the realignment of sediment grains due to gradually increasing stress resulting from increasingly deep burial). Sediment compaction has long been recognized as a major driver of subsidence in coastal Louisiana (Penland and Ramsey, 1990), but separating compaction from other subsidence processes is difficult. Recent studies have documented compaction rates of 5 to 10 mm/yr within the past decade (Jankowski et al., 2017) and 1 to 5 mm/yr within the past ~1,500 years (Törnqvist et al., 2008). Because compaction is so prominent in the shallowest subsurface (i.e., depths of 5–10 meters), it exhibits high spatial variability (Jankowski et al., 2017); as a result, rates of RSL rise can vary spatially by an order of magnitude.

Another potential driver of subsidence along the Gulf Coast is sediment loading. Isostatic adjustment due to sediment loading is a smaller contributor to subsidence (i.e., less than ~0.5 mm/yr) than glacial isostatic adjustment, even in a major area of sediment deposition such as the Mississippi Delta (Wolstencroft et al., 2014), and much smaller than sediment compaction. The spatial extent of sediment loading, however, is relatively large compared with shallow subsidence processes, and its footprint may extend well beyond the actual area of sediment deposition.

Finally, the withdrawal of subsurface fluids can play an important role in Gulf Coast subsidence. Fluid withdrawal can take place at multiple depths across the sedimentary column and a variety of fluids can be withdrawn (e.g., oil, gas, water, sulfur) for energy production, industrial uses, drinking water, cooling, and management practices. In urban areas such as New Orleans and Houston, groundwater is often pumped out as part of stormwater management actions. The resulting desiccation and increased oxidation rates of buried organic matter cause further subsidence (e.g., Snowden et al., 1980; Stephens et al., 1984; DeLaune et al., 1994). A prominent example of this is metropolitan New Orleans, where water withdrawal has caused parts of the city to subside by 3 meters or more in some places since about 1900 (Snowden et al., 1980; Dixon et al., 2006). Groundwater withdrawal has also been associated with up to 3 meters of subsidence in Galveston and Houston (Winslow and Doyel, 1954; Kasmarek, 2013). Groundwater is also removed from the subsurface for industrial purposes. Jones et al. (2016) used Interferometric Synthetic Aperture Radar (InSAR) to examine subsidence patterns in the greater New Orleans region for the period of 2009 to 2012 and found that areas of high, localized subsidence exist and are often associated with industrial groundwater withdrawal. Oil and gas withdrawal is also associated with subsidence elsewhere along the Gulf Coast. Spatial and temporal patterns in subsidence across Louisiana are correlated with hydrocarbon and water withdrawal and can be related to regional wetland loss (Morton et al., 2009; Morton and Bernier, 2010; Kolker et al., 2011). White and Tremblay (1995) suggest that oil and gas withdrawal in Texas is associated with regional patterns of subsidence and wetland loss, and that this withdrawal may have activated faults along the Texas Gulf Coast.

Episodic Coastal and Riverine Flooding

The Gulf of Mexico receives approximately 860 cubic kilometers (km^3) of water each year, including contributions from episodic riverine flood events. The Mississippi River system is the dominant source of freshwater to the Gulf, delivering about 534 km^3 of freshwater through the main stem of the Mississippi River and an additional 238 km^3 of freshwater through the Atchafalaya distributary. The next largest sources of freshwater are the Mobile (60 km^3), Apalachicola (20.5 km^3), and Sabine (13.1 km^3) Rivers. Moderate-sized rivers include the Brazos, Pascagoula, Pearl, Suwannee, and Trinity Rivers, which each discharge 7 km^3 to 10 km^3 per year (Dunn, 1996; Allison et al., 2012).

The amount of rainfall along the Gulf Coast can vary on seasonal, annual, and longer (e.g., decadal) timescales. In addition, episodic weather events such as tropical cyclones may produce much larger amounts of rainfall over hours than the total quantity accumulated over months prior to storm arrival. As climate change alters patterns of rainfall, the magnitude and timing of freshwater delivery to the Gulf Coast may also vary. Increased freshwater delivery during short time periods may result in substantial flooding in the coastal zone, particularly in highly urbanized settings with substantial hydrological alterations, especially if coincident with storm-induced surge.

The Gulf Coast is exposed to an average rate of three to four hurricane landfalls per decade (e.g., Doyle, 2009), frequently leading to inundation by storm surges (e.g., Needham and Keim, 2012). Because of the Gulf Coast's relatively gently sloping terrain, hurricanes generate significant storm surge when making landfall, with wind setup (i.e., water level rise due to momentum transfer by wind to the water column) being the dominant physical process. Forerunner surge inundation (i.e., arrival of high surge in advance of hurricane landfall), even if relatively small, will have more impact as sea level rises and could result in earlier inundation over greater areas and large, erosive waves over longer durations.

Eight of the top 10 most costly U.S. hurricane disasters occurred along the Gulf Coast (see Table 2.1). The very large spatial extents and intensities of Hurricanes Katrina and Ike led to widespread, high storm surge (Irish et al., 2008, Irish and Resio, 2010). Hurricane Katrina devastated the Gulf Coast, crippling the New Orleans metro region (e.g., Padgett et al., 2008; Pistrika and Jonkman, 2010) and dramatically altering the coastline of Louisiana and Mississippi (e.g., Day et al., 2007; Fritz et al., 2007). Hurricane Ike, with its large forerunner surge (Kennedy et al., 2011), caused widespread flooding and extensive degradation of the barrier islands in the Houston and Galveston metro regions (e.g., Sherman et al., 2013) and the Bolivar Peninsula. In 2004, Hurricane Ivan made landfall near Gulf Shores, Alabama, eroding beaches, damaging or destroying thousands of homes, and overwashing major roadways throughout Alabama and Florida's panhandle (Stewart, 2004). Flooding in inland areas is generally controlled by the balance between storm surge and freshwater drainage. During Hurricane Harvey, some areas of coastal Texas were inundated with storm surge, but the record-breaking damage and inundation in the Houston metropolitan area was dominated by rainfall-related flooding. Later that same year, Hurricane Irma caused widespread damage across Florida.

The exacerbation of hurricane impacts along the Gulf Coast by the combined influence of sea level rise and urban growth cannot be overlooked. RSL rise and substantial wetland

TABLE 2.1 Mainland U.S. Tropical Cyclones Causing at Least $1 Billion of Damage (not adjusted for inflation)

Rank	Tropical Cyclone	Year	Category	Damage (U.S.)
1	Katrina (southeast Florida, Louisiana, Mississippi)	2005	3	$125,000,000,000
1	Harvey (Texas, Louisiana)	2017	4	$125,000,000,000
3	Maria (Puerto Rico, U.S. Virgin Islands)	2017	4	$90,000,000,000
4	Sandy (Mid-Atlantic & northeast United States)	2012	1	$65,000,000,000
5	Irma (Florida)	2017	4	$50,000,000,000
6	Ike (Texas, Louisiana)	2008	2	$30,000,000,000
7	Andrew (southeast Florida, Louisiana)	1992	5	$27,000,000,000
8	Ivan (Alabama, northwest Florida)	2004	3	$20,500,000,000
9	Wilma (south Florida)	2005	3	$19,000,000,000
10	Rita (southwest Louisiana, north Texas)	2005	3	$18,500,000,000

NOTE: Katrina also caused damage to Alabama, but it was not included in this reference. SOURCE: National Hurricane Center, NOAA, 2018.

loss over the past century (e.g., Day et al., 2007) are estimated to have increased Hurricane Katrina's surge height by 1.3 meters (Irish et al., 2014). Yet, the impact of local sea level rise and long-term changes in land cover and topography on hurricane-driven inundation is complex and difficult to predict with confidence (Bilskie et al., 2014). Such impacts vary depending on geographic location and storm track characteristics (Atkinson et al., 2012).

RSL rise and coastal change, coupled with human development, will be the leading driver of hurricane flood hazard acceleration along low-lying sedimentary coasts such as the Gulf Coast (Woodruff et al., 2013; Reguero et al., 2018). Given future changes in storm climatology and mean sea level—but ignoring sea level rise's influence on long-term coastal change—several studies project a dramatic increase in hazard exposure. An example from outside the Gulf is the New York metro region, where flood elevations over a range of annual exceedance probabilities (e.g., 1 percent chance in a given year) are predicted to increase proportionally to the increase in sea level; while changes in storm climatology are predicted to only marginally increase the flood hazard (Lin et al., 2012). In areas farther inland, however, the influence of sea level rise on flood elevation may not be well captured by linearly adding present-day hazard curves to projected sea level rise. For example, the predicted relative increase in surge hazard in the bays of Panama City, Florida, is not linear with sea level rise. It either increases more slowly (at a rate as little as 85 percent) or more quickly (at a rate as much as 115 percent) than sea level rise, depending on the location (Taylor et al., 2015).

Levees, floodwalls, storm gates, and other infrastructure designed to reduce storm damage in specific geographic areas might impact storm surge and create damage in other nearby areas. When surge is prevented from entering the area landward of such a structure, the blocked water either then increases surge locally along the seaward side of the structure or will be redirected elsewhere. Often the response is for water to collect outside the area of protective infrastructure, increasing flood risks for homes and urban structures in that location. For example, the Louisiana *2017 Coastal Master Plan* shows an increase in 100-year flood storm surges 50 years into the future for areas outside of the levees and other hard infrastructure (see Figure 2.6; LACPRA, 2017). Across the Gulf Coast, new levees are in various stages of planning, construction, and completion. Notable examples include the Morganza to the Gulf levee (USACE, 2013b), a 98 mile structure that is nearing completion of its initial alignment (but which has not yet been built to 100-year standards), and the extended Galveston area coastal barrier (often called the "Ike Dike"), which is in the proposal stage but has not yet been federally authorized or funded (USACE, 2015).

Riverine Sediment Transport

The Mississippi River system is the dominant source of sediment for the Gulf system. It delivers about 136×10^6 metric tons of sediment per year (MT/yr), with about 88×10^6 MT/yr

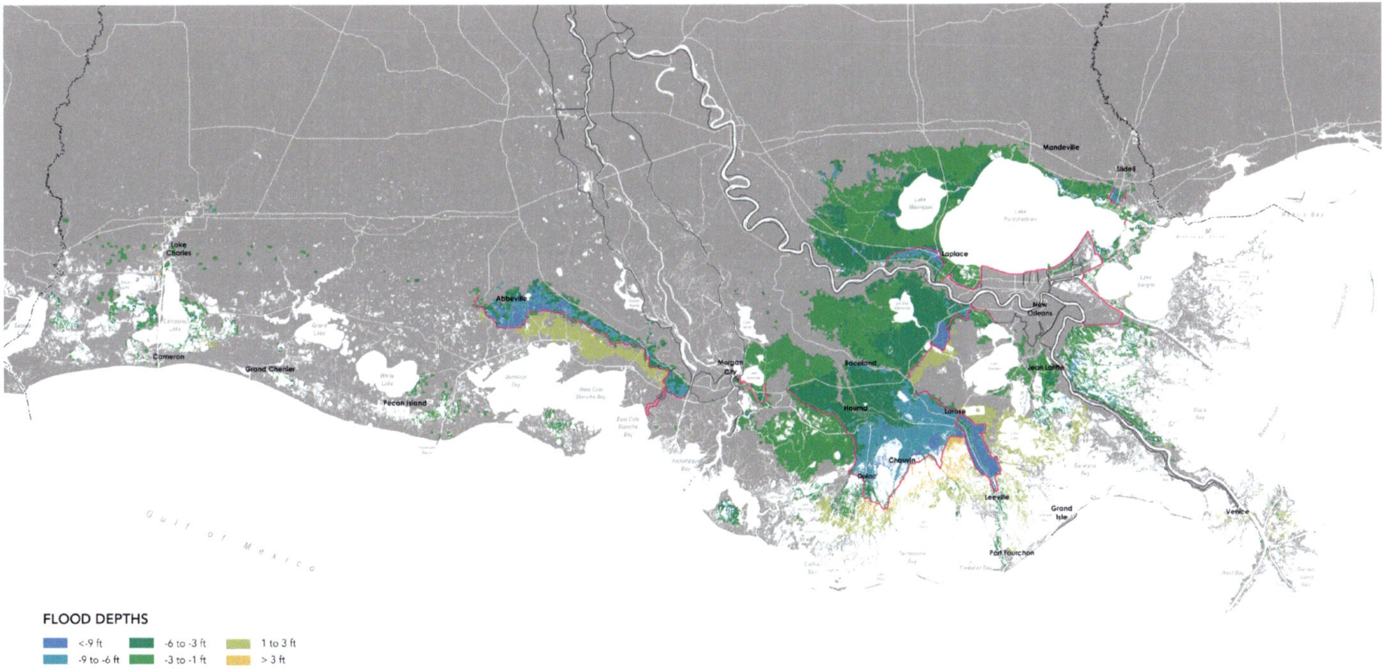

FIGURE 2.6 Changes in 100-year flood risk in the next 50 years in Louisiana with proposed master plan structural risk-reduction projects in place. Areas highlighted in yellow show induced flooding in front of proposed levees and floodwalls (pink lines). SOURCE: LACPRA, 2017.

coming through the main stem and 48×10^6 MT/yr through the Atchafalaya. The next largest sources of sediment are the Brazos (9×10^6 MT/yr) and the Mobile (4.5×10^6 MT/yr) Rivers. The Colorado, Pascagoula, Pearl, Sabine, and Trinity Rivers each deliver about 1×10^6 MT/yr of sediment (Allison et al., 2012). Coastal environments west of Mobile Bay, Alabama, are typically dominated by turbid waters, while coastal environments to its east typically feature clearer waters (Herrera-Silveira and Morales-Ojeda, 2009; Dzwonkowski et al., 2011). These differences stem, in part, from the nature of the river drainage basins. The Mississippi and Atchafalaya Rivers have a continental scale drainage basin that drains approximately 41 percent of the United States and contains diverse habitats including forests, farmland, arid lands, and cities; whereas many other rivers have regional-scale drainage basins with more homogenous land uses (Alexander et al., 2012). Because of their large absolute magnitude, the central Gulf rivers are important for both sediment and freshwater inputs (Horowitz et al., 2001; Turner et al., 2007; Allison et al., 2012; Lehrter et al., 2012; Fitzpatrick et al., 2017). Sediment budgets for the Mississippi River suggest that more than half of the sediment that enters at Tarbert Landing along the Mississippi–Louisiana border is trapped in the river channel or is transported to distributary channels near the river (Allison et al., 2012).

Coastal Sediment Transport and Hydrodynamics

Sediment sources within the Gulf of Mexico include offshore deposits from river discharge during periods of lower sea level and earlier periods in the Holocene Epoch, longshore drift from modern rivers, subsidence of coastal landforms, and biogenic production (Khalil et al., 2018). Sediments are available to be transported landward during normal coastal processes and during storm and overwash events and can be dredged for projects such as coastal restoration or mitigation of coastal erosion (Morang, 2006; Hickey et al., 2010; Wallace et al., 2010). These sediments are reworked by coastal hydrodynamics arising from tides, waves, and flooding during episodic coastal storms. The proportion of sediment trapped on coastal or deltaic plains matters greatly to the continued existence of coastal landforms and wetlands, but can be quite variable (sediment-retention efficiencies are commonly in the ~30 percent to 70 percent range; Blum and Roberts, 2009). Esposito et al. (2017) showed that within the Mississippi Delta, sediment-retention efficiencies can vary dramatically, with values of 5 percent to 30 percent near the open coast as opposed to greater than 75 percent in more inland settings.

Most of the Texas shoreline is characterized by a mixed tide regime, while in Alabama, Louisiana, and Mississippi the tides are diurnal. The west Florida shoreline has semi-diurnal tides in the northern part of the state and mixed tides in the southern part (Eleuterius and Beaugez, 1979). With rising sea levels, tides will be propagating on deeper water and tide components will change slightly. Pickering et al. (2017) project that, with a 2 meter sea level rise and no shoreline retreat, tidal amplitudes will be up to 10 percent higher than today on the west coast of Florida and west of the Mississippi River Delta. With shoreline recession, the tidal amplitude changes will be less.

Waves are the principle driving force for coastal processes. The Gulf of Mexico is a low-energy coastline, with nonstorm wave heights on the order of 0.3 meters.[4] Mean significant wave heights are larger in the eastern portions of the Gulf Coast (up to 2 meters), while maximum significant wave heights (associated with hurricanes) can reach up to 20 meters (Appendini et al., 2014). Results based on a 30-year hindcast (Appendini et al., 2014) indicate that the largest wave heights over this timeframe occurred offshore of Louisiana. Peak wave period values along the Gulf are generally 5 to 8 seconds, with the longer periods occurring in the eastern Gulf (Appendini et al., 2014). Panchang et al. (2013) showed, using a combination of wave modeling and buoy data, that 100-year return period values for deep water significant wave height on the order of 20 meters were appropriate. They also showed an increase in maximum annual significant wave height with time in the eastern part of the Gulf. Appendini et al. (2014) reached different conclusions in observable changes in wave height with time, with these discrepancies likely due to the use of small datasets in both studies (~50 and ~30 years, respectively).

Evolution of Coastal Landforms and Embayments

The geological record indicates that Gulf Coast bays and estuaries have a history of rapid and dramatic change in response to changes in sediment supply and RSL rise. Bayhead deltas, often found within coastal bays, are particularly sensitive to the balance between sea level rise and sediment input. When sediment input is sufficient, these deltas may grow despite rapid rates of RSL rise (e.g., Wellner et al., 2005; Anderson and Rodriguez, 2008). The Wax Lake (Louisiana) and Atchafalaya River Deltas, with their large sediment supplies, are among the few systems along the northern Gulf Coast that are likely to maintain or prograde during the next 50 years (LACPRA, 2017). Systems such as the Trinity River (within the Galveston Bay estuary complex) or Mobile River Deltas, with substantially less sediment supply, are likely to see land loss in the long term (Weston, 2014).

Many of the estuaries and bays of the Gulf Coast are fronted by sandy barrier islands and peninsulas. The remainder of the ocean coastline consists of marsh, bluff, and sandy mainland beaches. As sea level continues to rise, sandy coastline response will depend largely on the balance between the rates of sediment supply and RSL rise. When there is a sediment deficit, the shoreline will erode; when rates are balanced, shoreline position may remain stable over the long term. In the case of barrier island or sandy peninsula shorelines, when the rate of RSL rise outpaces rates of sediment supply, shoreline erosion can lead to

[4] See https://coastal.er.usgs.gov/coastal-classification/process.html.

landform narrowing. If sea level rise continues to outpace rates of sediment supply, landward migration will likely occur when barrier islands or peninsulas become narrow enough for storms to impact their entire width (Leatherman, 1979). Migration is facilitated by overwash processes, which occur when the combination of tides, storm surge, and wave action brings the water level above the height of the frontal dune or berm (e.g., Sallenger, 2000). Overwash carries sand from the front of a barrier or peninsula to the interior of the landform. In this way, storms promote the maintenance of islands and sandy peninsulas by building them upward and moving them landward, thereby maintaining island elevation relative to sea level as sea level rises. Barriers constructed from muddy deltaic deposits require a greater amount of landward migration to liberate the same amount of sand, compared with locations where the substrate contains a higher proportion of sand (Moore et al., 2010). As a result, deltaic barriers are more vulnerable to disintegration than barriers underlain by sand (e.g., Moore et al., 2010; Otvos and Carter, 2013; Moore et al., 2014). In general, shoreline erosion rates are highest (up to 78.6 m/yr) in Louisiana along barrier island and headland shorelines associated with the Mississippi Delta (Morton et al., 2004, 2005). High rates of erosion are also prevalent along some of the Texas barrier islands and headlands (up to 25 m/yr; Morton et al., 2004; see also Gibeaut et al., 2000; Wallace and Anderson, 2013), although some long beach segments in Texas have accreted due to the convergence of net longshore drift and the presence of tidal inlets that have been stabilized by long jetties. In contrast, barrier islands in Mississippi are migrating laterally (Morton, 2008; Otvos and Carter, 2008) due to net alongshore sediment transport rates in this region.

Interactions between vegetation and sediment transport cause localized sand deposition in the presence of beach grasses, giving rise to coastal dunes. When present, dunes may provide protection to inland habitats and coastal infrastructure by reducing vulnerability to overwash and flooding during storms. For this reason, dunes are sometimes constructed as part of beach nourishment efforts designed to provide recreational beaches and storm protection. Other coastal management strategies, including seawall construction or groin emplacement, are sometimes undertaken in an attempt to maintain shoreline position. Some of these strategies have been supplanted by beach nourishment efforts (Morton et al., 2004), likely because hard structures do not allow for landward migration and can lead to narrower beaches as erosion continues. They also block the transfer of overwash sediment to the back-beach and back-barrier environments, with potentially detrimental long-term effects. Other residential and commercial infrastructure can also have this effect, as demonstrated by a field study of several sites in New Jersey affected by Hurricane Sandy, where buildings and a boardwalk blocked the delivery of up to 90 percent of overwash sand (Rogers et al., 2015). A companion modeling effort suggests this has the potential to prevent barriers from building upward and migrating landward, thus hastening island disintegration or drowning as sea level rises (Rogers et al., 2015).

Barrier islands are most able to maintain elevation above sea level when they retain their mobility. Competition among processes that form dunes or cause erosion determine the likelihood that dunes will recover and therefore maintain their height (Durán and Moore, 2015). It is the spreading, herbaceous species such as dune grasses, rather than rigid and tall woody growth form, that promote sediment accretion and increases in elevation (Feagin et al., 2015). If coastal management plans focus instead on sediment stabilization and the attainment of later stages of vegetation succession, such as maritime forest, the system as a whole would lose its natural resiliency. When the substrate is no longer mobile, early successional stages disappear and colonizer species that are tolerant of burial under sand become locally extinct. As a result, biodiversity and the sand-binding function that works against erosion are diminished (Martínez et al., 2004). Roman and Nordstrom (1988) concluded that human-induced sediment starvation has profound effects on barrier island vegetation, including a threshold shoreline erosion rate beyond which vegetation does not recover. The exact magnitude of this threshold may vary among barrier islands and depends on shoreline orientation, tidal range, storm frequency, and the ability of native plants to bind sand (Roman and Nordstrom, 1988; Tsoar, 2005).

Forecasting the future behavior of the physical coastal system is challenging, but there are several ways that projections of future coastline position and landform state can be done. The first is to use the historical behavior of the shoreline or marshes, such as erosion rates over decades or vegetation growth rates, and to extrapolate these rates into the future. These empirical data-driven models have been used for short-term changes (i.e., years), as well as long-term (even geological) timescales. This procedure assumes that "what's past is prologue"; however, the recent past (i.e., the past 2,000 years) provides only a guide. For example, sea level is currently rising far faster than before, and so erosion rates in the future may be far higher than what has occurred in the past several thousand years.

Coastal evolution models, which simulate the evolution of the physical parts of the system, can be categorized by considering the domain across which a model operates and its underlying assumptions. For instance, coastal profile models only consider variations in water depth and topography in the onshore–offshore direction, either neglecting or parameterizing the effects of longshore variation (e.g., engineering models like SBEACH [e.g., Larson, 1990]; see Ruessink et al., 2007, for a review); while coastline models assume that the cross-shore beach profile is known and the same everywhere, but that the shoreline position can vary in the alongshore direction (e.g., Generalized Model for Simulating Shoreline Change [GENESIS; Hanson, 1989], Coastline Evolution Model [CEM; Ashton and Murray, 2006]; COastal Vector Evolution Model [COVE; Hurst et al., 2015], or see Dean and Dalrymple, 2002). Coastal area models allow the water depth to change in both the cross-shore and the alongshore direction (e.g., Delft3D [e.g., Lesser et al., 2004]; XBeach [Roelvink et al., 2009]); and landform models predict the evolution of barrier islands (e.g.,

Shoreline Translation Model [STM; Cowell et al., 1995], and those of Stolper et al., 2005; Masetti et al., 2008; and Lorenzo-Trueba and Ashton, 2014).

Coastal evolution models can also be classified according to whether they include all the relevant processes at the smallest scale practical (e.g., Delft3D) or include only those processes that are essential to capture realistic results (e.g., CEM, STM). Paola (2001) refers to these two types of models as reductionist or synthesist (i.e., reduced-complexity) models, while Bukulich (2013) uses the terms predictive and explanatory, respectively. Because synthesist models are sometimes used in a predictive sense, the Paola (2001) framework is used here. Reductionist models explicitly account for the bathymetric evolution via sediment transport formulations that directly link the mobilizing and transport of sediment to hydrodynamics. These models are often used for short-term events, such as bar migration or barrier island evolution, over timescales of days to weeks. The synthesist models are based on reduced-complexity formulations developed by choosing macroscale variables to drive the model, such as sediment transport rates, which inherently consider all the sediment transport modes and physics, but use simplified relationships instead of explicitly calculating them as a function of the overlying hydrodynamics. Even further abstractions are possible: Pape et al. (2010) use neural networks to compute the location and elevations of sand bars, migrating onshore and offshore. Reduced-complexity models can produce qualitatively reasonable behavior and some can be used to quickly explore parameter space within the context in support of engineering design, management planning, and decision making.

Ecological Processes

Gulf Coast ecosystems are highly diverse and provide essential services that are key in sustaining human and wildlife populations (de Groot et al., 2010). The dynamics, structure, and function of these ecosystems are broadly influenced by sharp environmental gradients that extend from the western to the eastern Gulf Coast, as well as substantial temporal variability that operates at different scales, ranging from episodic to seasonal to inter-annual. Human modifications of the coastal landscape, as well as climate change, also affect coastal ecosystems. Given the many interactions that can occur among natural environmental gradients, temporal variability, human coastal modifications, and changing climate, the impacts of such interactions on Gulf Coast ecosystems can be quite complex.

There are a multitude of ecological habitats along the Gulf Coast, including submerged sediments, wetlands, woodlands, and xeric dunes (see Figure 2.7). Wetlands fringe the coastline at the interface between land and the open waters of bays, lagoons, and the Gulf, comprising marshes and mangroves that are adapted to varying climate, seawater flooding, and salinity levels. With higher tolerance for cold weather, marshes tend to dominate in the northern Gulf Coast. Mangroves are currently more prominent along southern Florida and Texas, although there has been a recent northward expansion of red and black man-

FIGURE 2.7 Ecological systems found in the coastal landscape of the Gulf of Mexico. Representative examples from the north-central Gulf (from the Chandeleur Islands to Pensacola Bay) are presented. (A) Marshes at Point aux Pines (photo by Ryan Moody); (B) Mangroves on the Chandeleur Islands (photo by Ryan Moody); (C) Horn Island, a barrier island (photo by Caitlin Wessel); (D) Seagrasses in Little Lagoon (photo by Ashley McDonald); (E) Oysters in Portersville Bay (photo by Rochelle Plutchak); and (F) bare sediment flats at Point aux Pines (photo by Jason Stutes).

groves in the Gulf Coast (Scheffel et al., 2013; Armitage et al., 2015), most likely due to climate warming and a decrease in the frequency and duration of freezes in the northern Gulf (Osland et al., 2013).

The nearshore submerged bottom landscape includes seagrass beds, oyster reefs, and bare sediment flats. Turtle grass (*Thalassia testudinum*) and manatee grass (*Syringodium filiforme*) are more typical in marine, high-salinity conditions, while shoal grass (*Halodule wrightii*) and widgeon grass (*Ruppia maritima*) tend to dominate in brackish, estuarine conditions (Handley et al., 2007). Seagrass beds tend to be less extensive and productive in the western Gulf Coast (i.e., west of Mobile Bay to Texas) (McDonald et al., 2016a), due to turbidity from greater amounts of suspended sediment discharge from rivers.

Oyster reefs constitute prominent and productive ecosystems in many locations along the Gulf Coast. The extent and productivity of oyster reefs are temporally and spatially variable (Kim et al., 2010) due to the highly dynamic nutrient, salinity, and hydrodynamic conditions along the coast. Bare sediment flats (populated with benthic microalgae) are another common feature of the benthic landscape, whether interspersed among seagrass beds and oyster reefs or extending across vast expanses. Sediment type and grain size depend on the proximity of river inputs, geology of the region, and hydrodynamic processes, resulting in highly variable sediment mosaics. Due to the greater sediment loads, muddier sediments are often found in the western Gulf Coast (Bianchi et al., 1998).

Barrier islands support other important habitats along the Gulf Coast. The development of vegetation and geomorphic features may be more tightly coupled on barrier islands than in nearly any other ecosystem (Oosting, 1954; Godfrey et al., 1979; Stallins and Parker, 2003; Feagin et al., 2005; Durán and Moore, 2015). Plant species found on barrier islands are adapted to a broad range of stressors, including drought, salt spray, and freshwater or saltwater flooding (Oosting, 1954; Lee and Ignaciuk, 1985; Shao et al., 1996). The degree of shelter from these stressors provided by dunes and other landforms, along with soil characteristics, depth to the water table, and groundwater salinity, determine the composition and spatial distribution of barrier island ecosystems (Hayden et al., 1995). Exposure to freshwater or saline water is determined by the influences of surface elevation and topography on precipitation, infiltration, and runoff, and the likelihood of tidal (i.e., saltwater) flooding (Hayden et al., 1995).

Gulf Coast ecosystems provide habitats for a wide variety of fauna, including shellfish, finfish, waterfowl, and marine mammals, many of which have commercial, recreational, or conservation interest. For instance, many fish species spend their early lives in wetlands, seagrass beds, oyster reefs, and sediment flats, and upon maturity migrate to deeper coastal waters, where they are harvested by commercial and recreational fishing (Beck et al., 2001). More than 90 percent of all commercial fish landings in the Gulf of Mexico are species that use coastal shallow systems during some portion of their life stage (Kennish, 1999; Lellis-Dibble et al., 2008). Coastal habitats are also hotspots for many resident and migratory species of waterfowl, many of them protected or managed for hunting, including the endangered West Indian manatee (*Trichechus manatus*) (Hieb et al., 2017).

Seagrass beds, oyster reefs, wetlands, and barrier islands have the potential to reduce wave energy and help protect shorelines from erosion (Shepard et al., 2011). Wave-energy reduction also leads to enhanced sediment deposition, which contributes to shoreline accretion and protection (Manis et al., 2015). This is most evident when comparing storm impacts among coastlines with or without the protective effect of barrier islands, and also when comparing developed coastlands with no buffering vegetation in place to landscapes with buffering vegetation between the developed land and the coastline. Most often, storm impacts on land are more prominent with the absence of barrier islands and buffering vegetation (Woodruff et al., 2013; Sutton-Grier et al., 2015).

Coastal systems can also filter nutrients, preserving water quality. Through plant uptake and bacterial denitrification, wetlands can remove much of the nitrogen and phosphorus in land-derived runoff before it enters coastal waters (Sparks et al., 2015). Oyster reefs filter large amounts of particles out of the water column, enhancing water quality and clarity (Dame, 1993). Substantial nutrient filtration may also occur as terrestrial groundwater crosses the sediment before reaching the water column (Santos et al., 2012). Wetlands and seagrass beds can also act as large carbon sinks, since a portion of the carbon fixed through photosynthesis is stored as recalcitrant carbon (Macreadie et al., 2014; Hunter et al., 2015).

Impacts of Coastal Development and Climate Change on Ecosystem Dynamics and Function

The rapid expansion of human populations throughout the Gulf Coast during the past century has caused large changes in the coastal landscape (Jackson et al., 2001). Natural vegetation, such as maritime forests and wetlands, has been replaced with farmlands, residential areas, industrial facilities, and other urban structures (Valiela, 2006). Urban development increases the area of impervious surfaces in the watershed, leading to larger surface runoff to coastal waters. Farmlands and residential areas contribute to increased nutrient loading from fertilizers and septic systems. As an example, nutrient inputs due to human development may vary five-fold across coastal watersheds in western Florida (Christiaen et al., 2016; Stutes et al., 2017). Canals also can contribute to saltwater intrusion and wetland loss (Penland et al., 2000).

Taken together, these human-derived inputs may lead to pollution in coastal waters. This can result in fish kills and other losses of marine life (Cloern, 2001) and may also increase the turbidity of coastal waters, which in combination with dredging and boating activities (e.g., damage caused by grounding and propellers), may cause losses of seagrass beds (Hauxwell et al., 2003; Anton et al., 2011). Several studies in the Gulf have documented large seagrass losses following intense nutrient loading (Anton et al., 2011;

Stutes et al., 2017). Such nutrient loading can also be exacerbated by boating activities (Trevathan-Tackett et al., 2018). Additionally, hydrological modifications in the watershed can drastically affect freshwater flow into coastal waters, leading to salinity and sediment alterations that are damaging to marine life. Among the deleterious impacts of hydrological alterations by humans in the Gulf are decreases in system productivity, wetland erosion, and habitat loss (Beck et al., 2011; Garrote-Moreno et al., 2014; Sharma et al., 2016c).

Resource exploitation constitutes another human impact on coastal ecosystems. Historically, many Gulf Coast fisheries were subject to overfishing; affected species included red drum, red snapper, boccacio, and several species of groupers (Coleman et al., 2000, 2004; Cowan et al., 2011). The 2007 reauthorization of the Magnuson–Stevens Fishery Conservation and Management Act required strictly enforced annual catch limits and rebuilding plans for overfished stocks in federal waters. As a result, red snapper, gag grouper, hogfish, greater amberjack, and gray triggerfish were considered rebuilt and removed from overfishing lists, and overfished stocks decreased from more than 33 percent to less than 5 percent (NOAA, 2017). Many commercially and recreationally fished species, however, do not have fishery management plans and are not assessed (NOAA, 2017). Moreover, rules for state-managed fisheries vary by state and may put some stocks at risk of overfishing.

Climate change is another driver of change in coastal ecosystems. Increasing temperatures are shifting the ranges of occurrence for several species—for example, tropical species such as mangroves (Osland et al., 2013), manatees (Pabody et al., 2009), and surgeon and parrot fishes (Fodrie et al., 2010) are expanding northward to the northern Gulf due to warmer temperatures. Acidification, through enhanced carbonate dissolution and depressed carbonate precipitation, can affect the health of calcifying organisms (Lemasson et al., 2017), which are abundant in many areas of the Gulf Coast. Altered rainfall patterns through changes in the salinity of coastal waters can cause large changes in coastal ecosystem structure (Christiaen et al., 2016). Increased frequency of extreme weather events, including major storms and hurricanes, may also substantially alter the dynamics of coastal ecosystems (Chen et al., 2017). Rising sea level will be a particularly important driver of coastal ecosystem change in the years ahead. High rates of RSL rise will lead to landward transgression of coastal wetlands, provided there is space available for upland wetland migration (i.e., no "coastal squeezing"). If sediment supply is not adequate to maintain their elevation, wetlands will drown (e.g., Morris et al., 2002; Kirwan et al., 2010; Jankowski et al., 2017). There is evidence that sediment discharge from major rivers is declining, which will likely lead to decreased marsh accretion rates. Coupled with increases in sea level rise, this decline will put marshes at increased risk of succumbing to rising seas in many areas of the northern Gulf (Weston, 2014).

Sea level rise will also affect barrier island ecosystems. With a rise in sea level, depth to the water table is reduced. This reduction creates wetter habitats at the expense of drier ones, and profoundly changes species diversity, community structure, and ecosystem function

(Ehrenfeld, 1990). Ehrenfeld (1990) noted that the effect of sea level rise on barrier island vegetation had not been studied. Over the nearly three decades since, and facilitated greatly by the development and increasing availability of the global positioning system (GPS) and light detection and ranging (LIDAR) technology for measuring surface elevation (Gibeaut et al., 2003), substantial gains have been made in understanding this phenomenon. Recent studies in the northern Gulf Coast show changes in the predominance of barrier island habitats from drier to wetter plant community types over the past several decades (Lucas and Carter, 2010; Jeter and Carter, 2016). Critically, these changes are highly sensitive to even decimeter-scale changes in island elevation above sea level (Lucas and Carter, 2013; Anderson et al., 2016; Funderburk et al., 2016).

Effects of Strategic Natural Resource Conservation and Restoration on Coastal Ecosystems

Humans can alter coastal ecosystems through development of the coastal landscape, exploitation of marine resources, and climate change, with potential consequences for the ecosystem services on which humans rely. These changes point to the need for policies to promote sustainable development, with a goal of achieving environmentally and economically "smart" growth that preserves ecosystem service provision. Smart growth policies aim for vibrant economies and healthy environments.

Part of effective management is strategic natural resource conservation. Coastal development will inevitably cause some loss of ecosystem function, which can be minimized by conserving ecosystem functional hotspots that can help regenerate resources where they have been lost or grow those resources elsewhere in the developed coastal environment (McDonald et al., 2016b). Functional hotspots can regenerate or implant resources through larval production and dispersal (Kim et al., 2010), export of plant seeds that will subsequently settle (Kendall et al., 2004), and migration of adult individuals that can then reproduce in the colonized areas (Sala et al., 2002). For instance, a fringing marsh buffer may preserve many of the ecosystem services present in the predevelopment environment, such as filtering runoff (e.g., Sparks et al., 2015), providing habitat for juvenile fish (Moody et al., 2013; McDonald et al., 2016b), and attenuating wave energy to reduce the impacts of storm surge on nearby communities (Roland and Douglas, 2005). The preservation and use of marsh buffers in green and hybrid infrastructure approaches has recently gained interest (Sutton-Grier et al., 2015). Engineering in ways that make use of nature's ecosystem services and allow the natural system to adjust to changing conditions (e.g., by allowing marsh buffers to shift through time) is a main tenant of the emerging, innovative "Build with Nature" approach (e.g., de Vriend et al., 2015).

Coastal restoration, defined as the recovery of ecosystem structure and functions where they have been lost (Sparks et al., 2013; La Peyre et al., 2014), is another aspect of effective management. Coastal restoration can be achieved through various strategies, including land building or ecosystem creation. Restoration activities are now standard in many environmental management plans at local, county, and regional levels. In Louisiana, marsh creation and barrier island restoration occur through dredging and sediment pipelining, and partial diversion of the Mississippi River is being used to restart natural land-building processes. Restoration of degraded ecosystems can, at least to some extent, improve environmental health and help promote resilience in developed coasts (Christiaen et al., 2013; Sharma et al., 2016b). Restoration actions also have the potential to help reduce coastal hazards, such as storm damage and nutrient pollution. However, the success of restoration efforts is highly variable, with some projects performing poorly and others performing quite well (Sparks et al., 2013).

THE HUMAN SYSTEM

Although the scientific challenges of projecting physical and ecological changes, such as sea level rise or marsh loss, over the next 10 to 200 years are formidable, and sizable uncertainty is expected, the scientific basis to predict most significant human processes with any degree of confidence on these timescales is lacking. To provide some context for the difficulty in understanding human processes over the long run, some of the changes in U.S. society and the Gulf Coast over the past century are discussed below.

By 1910, the U.S. population was 92.2 million,[5] and the average U.S. life expectancy in 1910 was still far below modern levels, at 50.0 years.[6] New Orleans was no longer among the 10 largest urban centers in the United States.[7] The population of Florida was just 752,619 (compared with almost 21 million in 2017).[8] There was a small but emerging oil and gas industry in the Gulf Coast (Theriot, 2014), the tourism industry was nascent, and the seafood industry was dramatically smaller. From 1950 to 2016, total commercial fishery landings in the Gulf Coast grew by a factor of three, from 258,841 MT to 791,433 MT.[9] Over the same time span, total U.S. commercial fishery landings only grew by a factor of two. Gross domestic product per capita in the United States more than tripled, from $13,819 in 1950 to $51,338 in 2016 (in 2009 dollars).[10]

While quantitative changes in the size of the population, life expectancy, the U.S. economy, and Gulf Coast industries have been massive over the past century, qualitative changes in the nature of economic activity and how people live are arguably even more substantial

[5] Data available at: https://www.census.gov/history/www/through_the_decades/fast_facts/1910_fast_facts.html.

[6] Data available at: https://data.cdc.gov/NCHS/NCHS-Death-rates-and-life-expectancy-at-birth/w9j2-ggv5.

[7] Data available at: https://www.census.gov/population/www/documentation/twps0027/tab07.txt.

[8] Data available at: https://www.census.gov/dmd/www/resapport/states/florida.pdf and https://www.census.gov/quickfacts/FL.

[9] Data available at: https://www.st.nmfs.noaa.gov/commercial-fisheries.

[10] Data available at: https://fred.stlouisfed.org/series/A939RX0Q048SBEA.

and would have been impossible to predict. For example, the 20th century saw electrification of rural areas; widespread adoption of interior plumbing in rural areas; improvements in safe drinking water and sanitation in both rural and urban areas; widespread adoption of air conditioning (especially important for Gulf states); radical transformations in transportation systems with the emergence of aviation and the ubiquitous use of the automobile; major advances in medicine including the development of antibiotics; the emergence of the computer industry; and revolutions in both communications technology and media (Gordon, 2016). These examples are well known and are mentioned as a humble reminder of how difficult it is to anticipate what human society can or will achieve over something as large as a century timescale.

There is an argument, however, that many of the most important changes in the 20th century are unique to that time, and changes like them can never occur again. Economist Robert Gordon argued that growth in the U.S. economy, including the Gulf Coast, in the 20th century was largely attributable to qualitative changes and adoption of technologies, such as the ones described above, that improved human well-being in ways that cannot be repeated in the future (Gordon, 2016). Moreover, population growth in the United States has slowed substantially. It may not be entirely absurd to model human processes with the expectation that fundamental ways of living and basic construction, transportation, and infrastructure technologies will continue to be similar to today.

Many aspects of coastal storm damage or erosion reduction have remained unchanged over centuries. The basic design principles of these structures, such as breakwaters or jetties, date back centuries (e.g., Bruun, 1972; Charlier et al., 2005), though refinements and economies have been implemented in more recent times. Many other practices, such as beach nourishment, have been refined and documented, but the basic principles have remained the same. There have been some refinements—while streams have long been diverted, the scale at which diversions are being planned for land building in the Gulf is new (e.g., channels that carry 75,000 ft^3/s; LACPRA, 2017).

While not all coastal changes will have direct day-to-day impacts, many of them will require some adjustment in management strategies and decision making. There are two primary pathways through which humans can respond to coastal change: they can adapt in place or they can migrate. Adapting in place might entail defensive capital expenditures (e.g., government and household spending on durable goods for climate change adaptation), restrictions on development, or simply choosing to do nothing. Relevant decisions may be made at the federal level (e.g., by the U.S. Army Corps of Engineers [USACE], the Federal Emergency Management Agency, the National Flood Insurance Program), the state level, locally at the town or municipal level (e.g., floating municipal bonds to fund a nourishment project), by individual households (e.g., installing revetments and bulkheads), or jointly by combinations of these groups. For example, the management of public flood risk has long focused on large-scale engineering projects, such as seawalls and levees, designed and implemented by government agencies (Martin et al., 2010; USACE, 2014). An important Gulf Coast example is the Greater New Orleans Hurricane and Storm Damage Risk Reduction System, which is intended to prevent or reduce flooding from storm surge and waves from tropical cyclones in New Orleans. After the system catastrophically failed during Hurricane Katrina, USACE upgraded the system at a cost of approximately $14.5 billion (USACE, 2012). Much of the coastal flood control infrastructure built by USACE, in fact, is located in Louisiana and Texas (NRC, 2013; USACE, 2018).

Recently, there has been a shift toward a more integrated approach, including flood prevention and damage alleviation through small-scale measures taken by communities and households, such as flood-protection devices (e.g., flood vents), adaptive building uses, elevating homes, and flood insurance (McDaniels et al., 1999; Samuels et al., 2006; Interagency Climate Change Adaptation Task Force, 2011; LACPRA, 2017). The success of these programs depends on local residents' willingness and ability to undertake those measures.

Adaptation Decision Making

Of the Gulf Coast states, only Louisiana and Texas have developed some version of a coastal adaptation plan. Louisiana recently finalized its third 50-year *Coastal Master Plan* (LACPRA, 2017), which provides information to stakeholders about the potential impacts of climate change and other drivers on Louisiana's coastal areas. The *Coastal Master Plan* also identifies and prioritizes risk-reduction and mitigation and restoration projects. Louisiana's coastal master planning process originates from the 2005 hurricane season, which included Hurricanes Katrina and Rita. The planning process is supported by one of the most sophisticated and integrated coastal simulation modeling efforts to date, with a series of linked models that allow the state to project future coastal changes, as well as potential mitigation and restoration scenarios.

Texas also recently issued its first *Coastal Resiliency Master Plan*, which provides information about the risks and impacts of climate change, as well as guidance for adaptation measures (Texas General Land Office, 2017). Both the Louisiana and Texas plans included the involvement of state and local constituents, as well as input from federal agencies, academic institutions, think tanks, and nongovernmental organizations. The effort in Louisiana invested significantly in policy analysis and public and stakeholder engagement (LACPRA, 2017; Speyrer and Gaharan, 2017). In addition, Florida, Louisiana, and Texas have adaptation planning activities at the local and the regional level.[11] Florida has initiatives in the counties of Broward, Lee, Miami-Dade, Pinellas, and Sarasota, while Texas has initiatives in the Austin and Galveston-Houston areas. Although not all states have adaptation plans designed to address the impacts of climate change, all do have some level

[11] Additional details can be found at: http://www.adaptationclearinghouse.org.

of coastal restoration activities that include elements of adaptation with implications for coastal resiliency (see NASEM, 2017b, for examples).

There is a growing understanding of how adaptation decisions are made. Individual adaptation measures have demonstrated ability to reduce flood damage (Kreibich et al., 2005; Schanze, 2008), yet relatively few people undertake adaptation measures voluntarily (Kunreuther, 1996; Wong-Parodi et al., 2017). It has been commonly assumed that individuals adapt as a result of high levels of perceived risk. A recent review, however, found that individuals' willingness and ability to take such measures was unrelated to their perceptions of flood risk (Bubeck et al., 2012). Bubeck et al. (2012) suggest that taking steps to reduce risk (i.e., adaptation measures) is associated with decreasing levels of perceived personal risk, even if absolute risk remains unchanged. They also found less willingness to act among individuals who estimated higher costs for these measures, preferred public flood defense measures, or saw government as responsible (see also Kellens et al., 2013). Conversely, and echoing findings with respect to climate change perceptions (Lee et al., 2015), the review found greater stated willingness to adopt individual adaptation measures among people who viewed them as effective, who knew more about flooding hazards, and who had experienced flooding directly.

Other factors may also be important for the adoption of adaptation measures, in terms of actually being able to follow through with the intent to take action (see Box 2.2). For example, some studies found that those who have greater social support (e.g., someone to call if they need help) are more likely to take protective measures before, during, and after disasters (Riad et al., 1999; Kaniasty, 2012). Other studies found, however, that people who have greater levels of social support are more likely to take risks than they would otherwise, as they can treat that support like a safety net (Weber and Hsee, 1998; Hsee and Weber, 1999; Schneider et al., 2013). A recent study found that among individuals impacted by Hurricane Sandy, those who reported greater social support also reported greater tolerance for flood risks and greater confidence in community adaptation measures, suggesting an important but complex role of personal connections to collective resilience both keeping people in place and helping them survive there (Wong-Parodi et al., 2017). These findings are likely generalizable to the Gulf Coast, where individuals face similar threats due to extreme weather events and geography. For example, a study conducted in Mississippi found that individual and community characteristics make a difference with respect to overall resilience. Sherrieb et al. (2010) found that greater median household income, greater education, and greater employment were associated with greater overall community resilience.

Government and Household Spending for Climate Change Adaptation

Long-lasting community and individual investments are often made to mitigate potential harms from environmental change. These investments are known as durable goods or

BOX 2.2
Examples of Adaptation Studies Looking at Human Behavior

There is growing interest in developing behavioral interventions to encourage the adoption of adaptation measures. A recent study conducted by Wong-Parodi et al. (2018) examines the predisposition to action and the impact of a sea level rise decision aid that emphasizes the risk, the adaptation measures, both, or neither on the adoption of adaptation measures among a representative sample of 1,201 at-risk coastal residents (living within a 100-year floodplain) in Connecticut, New Jersey, and New York. The researchers found that individuals who report having taken action previously are more responsive to decision aid messages, with the exception of the combined message (i.e., both risk and protective actions). The combined message had a positive effect on those who had not previously acted, but a negative effect on those who had. This result suggests that combining risk information with plausible adaptation strategies can enhance adaptation among most people, who do not typically take action. Among those who do take action, however, the decision aid may have generally reduced participants' estimates of flooding risk. The purchase of flood insurance is one type of adaptation measure that people can take to protect their homes. Landry and Jahan-Parvar (2011) found certain household characteristics, including greater household income and higher perceived risk, were associated with greater likelihood of having flood insurance in seven southeastern counties, including some in Texas and Florida. Michel-Kerjan and Kousky (2010) found policy holders tended to be single-family residential households and in flood-risk zones. The authors also found that people who do not experience a catastrophic loss tend to drop their policy over time.

In addition to the adoption of long-term adaptation measures, some studies have looked at shorter-term behaviors intended to protect property from damage due to immediate threats such as from a hurricane. In a recent study, Wong-Parodi and Feygina (under review in *Weather, Climate, and Society*) followed 521 Florida residents (including those living on the Gulf Coast) in the days leading up to and during Hurricane Matthew. They found that those who are within an evacuation zone, are informed by a first responder, and express greater acceptance of climate change are more likely to take measures to protect their homes. Moreover, those people with greater mental health and higher levels of self-efficacy respond in ways commensurate with the risks they face—they take protective measures if they are within an evacuation zone and do not if they are not at risk. However, those with lower levels of mental health and self-efficacy do not respond in ways that are commensurate with the risk—those with lower levels of mental health prepare when not at risk and those with lower levels of self-efficacy do not prepare when they are at risk. Finally, the researchers found that social support moderates the relationship between self-efficacy and protective behaviors, with the lowest and the highest levels of social support exhibiting greater self-efficacy.

defensive expenditures (e.g., Bunten and Kahn, 2017) and they tend to be large purchases. At the household level, a family may decide to invest in a truck with high ground clearance in order to avoid having their daily commute impacted by recurrent nuisance flooding. As another example, a household trying to mitigate flood risk may invest in interventions such as elevating the structure on pilings or installing hard, waterproof flooring instead of carpeting that is susceptible to mold. The Louisiana *Coastal Master Plan* specifically calls for these nonstructural interventions based on projected future 100-year flood depths: waterproofing for 1 to 3 feet, raising structures for 3 to 14 feet, and acquisition of the property (abandonment) for more than 14 feet (LACPRA, 2017). Such decisions may enable households to continue living in coastal areas that experience moderate environmental changes. However, the ability of individual households to invest in structure mitigation or other adaptation measures will vary greatly with financial capacity and knowledge to take action, and also depends on the availability of incentives to support such investments.

Governments also invest in durable goods, often in the form of defensive engineering structures, to adapt to climate change. For example, a beachfront municipality may choose to build a seawall or nourish their beach to mitigate against storm damages. Household decisions on durable goods for adaptation can theoretically feed back on the natural system as well. A group of houses on a barrier island that are raised on pilings could allow more overwash to occur compared with the same houses that were not elevated (e.g., Rogers et al., 2015). Nevertheless, the effects of large-scale government defensive expenditures may create more pronounced feedbacks. For example, the Lake Borgne Surge Barrier constructed after Hurricane Katrina was expected to alter hydrodynamics in ways that are relevant to navigation (Martin et al., 2010).

Key drivers of defensive expenditures in the coastal zone are (1) federal-, state-, and locally-funded projects for maintaining channels, dredging ports, stabilizing shorelines with seawalls and beach nourishment, and building levees; (2) federal-, state-, and locally-subsidized flood insurance; (3) ecological restoration and land-creation projects that are funded with the explicit intent of reducing storm surge impacts and restoring coastal habitat (LACPRA, 2017); and (4) investments made in elevating or flood proofing at-risk buildings. For example, all the Gulf states have undertaken beach nourishment projects using federal funding.[12] The extent to which federal, state, and local governments continue to contribute to these initiatives (and the potentially changing role of the private sector) is a key driver of which projects are built, development patterns in flood-prone areas, and how much local public finance will compensate by spending more of its local tax base on defensive investments. The possibility for important feedbacks implies that modeling the long-term evolution of the natural-human coastal system involves understanding when, where, to what extent, and in which forms defensive expenditures will occur (Smith et al., 2015; Gopalakirshnan et al., 2016).

Migration

The study of migration in response to coastal environmental change is a broad topic that touches on many academic fields, including demography (Hauer et al., 2016), anthro-

[12] Details about the Program for the Study of Developed Shorelines are available at: http://beachnourishment.wcu.edu.

pology (Oliver-Smith, 1996), human geography (Colten et al., 2008; McLeman, 2017), history (Gutmann and Field, 2010), psychology (Thompson et al., 2017), political science (Aldrich, 2012), sociology (Myers et al., 2008; Meyer et al., 2018), and economics (Bunten and Kahn, 2017). These various fields approach the problem with different theories, data, and methodologies. For example, recent work in demography uses past empirical patterns of migration to extrapolate future patterns in response to sea level rise without modeling the causal drivers of migration explicitly (Hauer, 2017). In contrast, work in history and human geography usually relies on detailed case histories of natural hazards and other environmental changes (McLeman and Smit, 2006; Colten et al., 2008; Gutmann and Field, 2010). Political scientists and economists tend to use statistical methods that relate relocation and migration decisions to underlying structural drivers (Timmins, 2007; Bayer et al., 2009; Aldrich, 2012).

Despite methodological differences, there is some agreement on the factors that influence migration. People move or stay in response to economic opportunities, physical and economic ability to migrate, preferences for public goods (e.g., environmental amenities, schools, cultural amenities), and social capital (e.g., family, friends, civic engagement, religious organizations, ethnic and language ties). In general, communities with greater social vulnerability are less able to migrate away from increasingly hazardous areas (e.g., Myers et al., 2008). One such example is the Isle de Jean Charles, a small island in Terrebonne Parish, Louisiana. Located outside the Morganza to the Gulf levee, the island has seen a loss of more than 98 percent of the community's land during the past 60 years, and only 320 acres of the island's original 22,400 acres remain (HUD, 2017). Over the past half-century, as residents left (largely due to hurricane or flood damage) and access to work and schools became restricted due to regular flooding of the road connecting the island to the mainland, the number of residences (many now elevated and protected by ring levees) has dwindled by more than 60 percent (HUD, 2017). Based on the historic trends of land loss and future projections of sea level rise, as well as damage from storms, the island will most likely not be habitable in a few decades. The community has partnered with the State of Louisiana to look at resettlement options, which will be funded by a $48.3 million grant awarded by the U.S. Department of Housing and Urban Development (HUD, 2017).

Energy-Related Infrastructure

The energy sector is characterized by a complex and diverse set of infrastructure components (i.e., social, economic, environmental, built) within a multifaceted operational environment. Energy infrastructure provides essential fuel to a multitude of dependent economic sectors. In turn, energy infrastructure depends on other sectors such as transportation, information technology, communications, financial services, human and environmental resources, and government facilities (DHS-DOE, 2015). An illustration of these complex

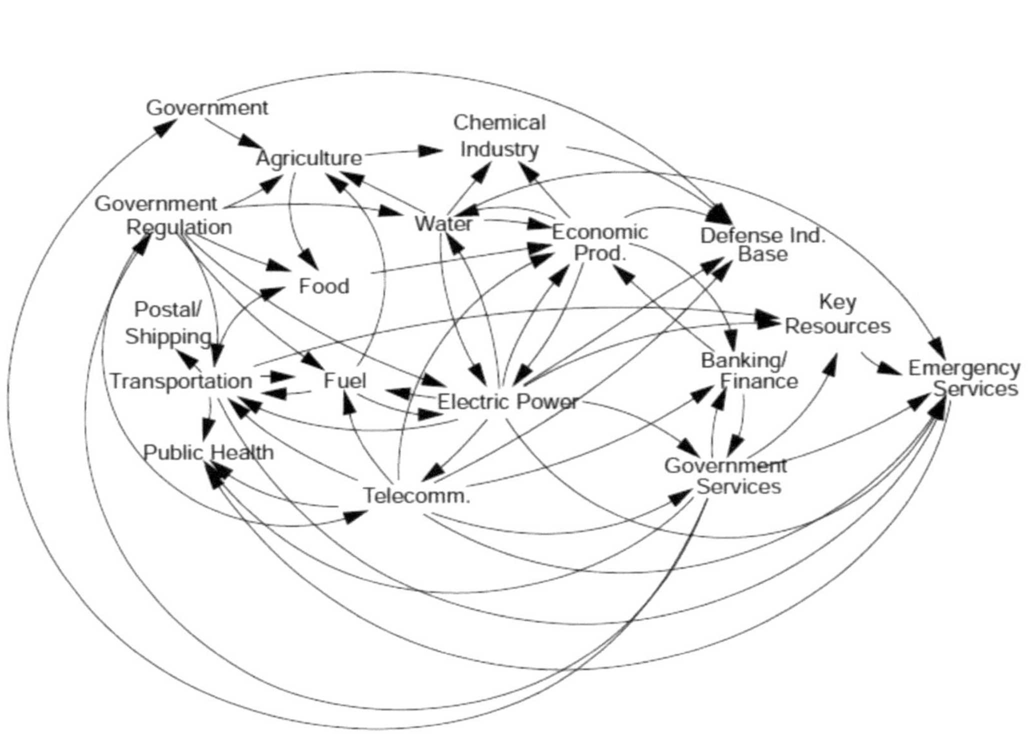

FIGURE 2.8 Illustration of infrastructure connections. SOURCE: Wilbanks and Fernandez, 2013.

connections (see Figure 2.8) demonstrates that a disruption in a single component can generate disturbances to other sectors. While the scale of the disturbance depends on the size of the disruption, the interconnections can amplify the effects of a disruption, both in terms of severity and extent, causing large cascading failures (Wilbanks and Fernandez, 2013). For example, as part of a study commissioned to assess the impacts of hurricanes on energy infrastructure along the Gulf Coast, Dismukes (2011, p. v) concluded that "a disruption in the region's infrastructure can have dramatic implications for not only domestic but also worldwide energy markets."

The American Society of Civil Engineers (ASCE) grades the current state of national energy infrastructure as "D+" due to a combination of aging and increased vulnerability to storm impacts (ASCE, 2017). For example, 70 percent of U.S. high-pressure natural gas pipelines were installed before 1980 (DOT, 2011; DOE, 2015). Much of this oil and gas pipeline network is along the Gulf Coast.

The geology and geography of the Gulf Coast has influenced access to resources and the technologies used to exploit them (Priest, 2007). For more than half a century, major refineries and petrochemical plants have been in operation in the region, albeit with process and capacity modifications, reconditioning, and upgrades along the way. Much of the fabrication and shipbuilding industry supporting oil and gas exploration is also more than a century old (BOEM, 2014). Age might not necessarily be directly indicative of vulnerability, but it is suggestive of older design practices that may be more vulnerable to changing future conditions. Infrastructure age often acts together with and may reinforce the effect of other factors such as design, maintenance, and operation in increasing the vulnerability of infrastructure to these various threats (DHS, 2010).

Additionally, the nation's strategic petroleum reserve, which stores more than 700 million barrels of oil, is located along the Gulf Coast, tapping into salt domes and other subsurface features at points along the Texas and Louisiana coasts. The four storage facilities are at risk from storm surge caused by sea level rise and hurricanes (USACE, 2013a; Bradbury et al., 2015). A recent assessment by the U.S. Department of Energy noted that much of the Gulf Coast energy infrastructure is past its design life (DOE, 2016).

Storm surge flooding is the predominant hazard responsible for damages to energy infrastructure during hurricane events (NIST, 2006; Harris and Wilson, 2008). Storm surge flooding impacts to energy infrastructure have occurred in the past when perimeter berms around facilities, intended to provide protection, were overtopped by storm surge that exceeded the berm elevations. Flood protection of structures (e.g., by floodwalls, levees, and berms) or raising the height of existing structures and elevating critical equipment and control systems above design flood elevations have been the primary hardening efforts undertaken by the industry (DOE, 2010). In terms of resiliency, the industry has invested in preparedness activities such as maintaining minimum tank volumes, staging command vehicles and portable generators, coordinating priority restoration of services, and planning for employee evacuation exemption and reentry (DOE, 2010). Another issue is external metal corrosion from salt water, which is especially destructive to energy infrastructure (DOE, 2010). For example, between 1994 and 1999, of the major (i.e., resulting in injury, fatality, or more than $50,000 in property damage) hazardous liquid and natural gas transmission pipeline accidents, 65 percent and 35 percent, respectively, were due to external corrosion (Beavers and Thompson, 2006).

Besides the infrastructure components previously discussed, there is a large group of ancillary but equally critical oil and gas infrastructure that interacts with coastal physical processes. These include fabrication yards for various oil and gas infrastructure (e.g., see Dismukes, 2011); water intakes and outfalls; facilities for wastewater and stormwater detention; facilities for management, disposal, construction, and operation of diverse products and materials (Barr, 2014); seawalls, revetment, levees, dikes, and wetlands for protection of facilities (van der Meer et al., 2008; DOE, 2010); and groins, breakwaters, natural materials, and rock and concrete armor for protection of pipelines and pumping stations (NRC, 1994; The Nature Conservancy, 2013). Similar to the larger main production and processing units, these types of infrastructure are also vulnerable to risks associated with the changing coastal system, especially to impacts from subsidence and sea level rise via changes in their relative elevation. In the central Gulf Coast region (between Galveston, Texas, and Mobile, Alabama), 72 percent of ports, 27 percent of major roads, and 9 percent of rail lines are at or below 4 feet of elevation (CCSP, 2008).

The Gulf Coast region is projected to remain an important contributor to crude oil and gas production in the world (API, 2017; EIA, 2018). In addition to increased core energy infrastructure that will be needed to support higher demand for oil and gas resources, supporting infrastructure critical to the supply chain along the Gulf Coast can also be expected to increase (API, 2017; ASCE, 2017). It is worth noting that recent analyses of energy cost and policy suggest that renewable energy growth across the United States is poised to continue (Breton and Moe, 2009; Lopez et al., 2012; NREL, 2016). While renewable energy resources and infrastructure are currently relatively scarce across the Gulf Coast, this could change over the next several decades.

In addition to energy infrastructure, the Gulf Coast is home to a wealth of other infrastructure assets, many of which could be affected by long-term physical changes to the coast. Some of this infrastructure is unique to the Gulf Coast and some is typical of many parts of the United States. Other critical infrastructure includes numerous ports; the Mississippi River, which is among the most important shipping pathways in the nation; and the Gulf Intracoastal Waterway, which connects inland waterways across the Gulf. The region has an extensive road and railway network, with some roads located at or near sea level. Some roads were built as causeways and others (e.g., the portion of LA-1 from Golden Meadow to Port Fourchon) were originally built as surface roads and have been elevated to provide an evacuation route from areas prone to flooding. An extensive inventory and description of Gulf Coast non-oil and gas infrastructure and its vulnerability to RSL rise and hurricane hazards is available (see CCSP, 2008; DOE, 2010, 2015; Wilbanks and Fernandez, 2013).

FEEDBACKS WITHIN THE COUPLED NATURAL-HUMAN SYSTEM

Especially important in feedbacks between the natural and human systems are the efforts undertaken to control coastal erosion and reduce flooding hazards through infrastructure approaches. Contemporary practice in coastal engineering for long-term and chronic coastal hazard mitigation includes consideration of traditional "gray" (hard) approaches (e.g., seawalls, bulkheads, groins) and "green" (nature-based) infrastructure (e.g., beaches, dunes, barrier islands, wetlands, mangroves, reefs), and the hybrid combination of the two, along with hazard mitigation investments in structure elevation, flood proofing, and buyouts. The design, performance, and impact of gray infrastructure in the coastal zone have been studied for some time (ASCE, 2003). Thus, the following discussion emphasizes green and hybrid infrastructure.

There is strong evidence that green infrastructure affects long-term coastal change—for example, by slowing chronic shoreline erosion and promoting sustainability of critical habitats—and plays a role in episodic inundation- and wave-driven processes (e.g., Sutton-Grier et al., 2015; Narayan et al., 2016)—for example, by damping waves and wave run-up (e.g., Mendez and Losada, 2004; Irish et al., 2014). This is certainly true with existing in-place green infrastructure. As with any engineered coastal structure, green infrastructure may at once serve to reduce the hazard (long term or episodic) in the intended design area and potentially lead to an increase in hazard in other areas (e.g., Wamsley et al., 2009), an unintended consequence. Green and "Build with Nature" approaches (e.g., de Vriend, 2015) are more likely than gray infrastructure to provide the benefit of other ecosystem services, which makes them possible alternatives in new and ongoing coastal resilience initiatives.

Hybrid approaches incorporate nature-based approaches (e.g., oyster reef emplacement or dune construction) to enhance the effectiveness of gray infrastructure (Sharma et al., 2016a). Besides helping human-made features become more efficient, hybrid approaches can also help preserve ecosystem function and services. One example is the use of marshes, revetments, and breakwaters to reduce wave energy and enhance coastline resilience against storm surge. Marshes may further protect shorelines from erosion when conserved or restored alongside gray infrastructure and can provide ecosystem functions such as fisheries productivity and filtration. As with any engineered erosion-control measure, care needs to be taken to minimize erosive impacts downdrift. In Louisiana, a hybrid "multiple lines of defense" approach has emerged. This strategy posits that a combination of natural features (e.g., marshes, barrier islands), built infrastructure (e.g., levees, floodwalls), and nonstructural measures (e.g., evacuation planning) are necessary to reduce the impacts of storm surges (Lopez, 2009). This concept has been partially incorporated into the development of Louisiana's Coastal Protection and Restoration Authority, which is responsible for coastal restoration and flood-risk reduction (LACPRA, 2017). See Box 2.3 for further discussion of feedbacks associated with gray and hybrid approaches.

Projecting the future evolution of the coupled natural-human system is particularly challenging as it requires both an understanding of, and the capability to simulate, the evolution of each system and its components, as well as the dynamics of feedbacks among all aspects of the system. There are recent examples of integrated simulation models used to project future conditions and support long-term coastal planning on the Gulf Coast that include consideration of feedbacks between the natural and human coastal systems. Specifically, models developed for Louisiana's 2012 and 2017 *Coastal Master Plan* (LACPRA, 2012, 2017; Peyronnin et al., 2013; Brown et al., 2017) encode a series of relevant relationships among coastal processes. Key relationships include, for example, feedbacks between coastal eco-hydrology and wetland morphology (Meselhe et al., 2013; White et al., 2017) or morphology and coastal vegetation (Visser et al., 2013). Interactions include influences from these drivers on either key habitat types or on the propagation of storm surge and waves during tropical storm events (Peyronnin et al., 2013; Brown et al., 2017). This sys-

BOX 2.3
Feedbacks Related to Coastal Protection Infrastructure: Examples from the Gulf Coast

A classic example of a self-reinforcing feedback is the emplacement of hard structures to protect coastal development, such as the engineering solutions undertaken after the 1900 hurricane devastated the city of Galveston. The emplacement of defensive engineering tends to lead to increases in coastal development and therefore the need for additional defensive engineering, leading to reductions in coastal erosion and prevention of island rollover. For example, the devastation following the 1900 hurricane led to a decision to raise the elevation of Galveston Island by 8 feet to 16 feet and to emplace a concrete seawall along part of the island (USACE, 1981; Hansen, 2007). Numerous modifications and extensions to the seawall, the beaches in front of the seawall, and the embankments behind it have been made since original construction, continuing to the present day (Songy, 2017). This change in the natural system significantly reduced vulnerability to future storms, preventing damage and loss of life during subsequent hurricanes, such as the 1915 hurricane and Hurricane Ike in 2008. These changes have also prevented the shoreline from eroding beyond the location of the seawall and continue to prevent the natural landward migration of the barrier island. Erosion of the landward beaches by storms led to a proposal for a groin field and artificial nourishment back in the 1930s; even then, a hybrid solution was envisioned. This hybrid approach continues to this day—a massive nourishment project on the western portion of the seawall was carried out in 2017 (Songy, 2017). By comparison, on nearby Bolivar Peninsula, which was not elevated following the 1900 storm and where the beach was backed only by low-lying dunes, the impacts of Hurricane Ike were catastrophic. Approximately 3,600 buildings out of 5,900 were destroyed (FEMA, 2009), the island was stripped of nearly all vegetation, and there was significant loss of life (Kennedy et al., 2011; Sherman et al., 2013). However, the lack of hard structures on Bolivar Peninsula allowed the landform to adjust to changing conditions, enhancing the likelihood that this feature (and the protection it provides for the bay and mainland behind it) will persist in the long term.

Similarly, construction of a shore-perpendicular groin or groin field that curbs erosion by trapping sediment also causes a self-reinforcing feedback. The potential for a groin in isolation to cause downdrift erosion is well recognized, so groins are generally designed as a groin field and used in conjunction with beach nourishment. When not properly designed, constructed, or maintained, however, groins can transfer the erosion problem downcoast or cause locally worsening erosion, continuing the cycle of a change in the physical system, which leads to further manipulation by humans and then further alteration to the physical system. At Longboat Key along Florida's Gulf Coast (see Figure), a groin field consisting of closely spaced groins without beach nourishment was ultimately removed in favor of a nourishment-only approach (Morton et al., 2004), but was later supplemented with the emplacement of new permeable groins (Harthill, 2014).

Another example of feedbacks related to coastal engineering approaches comes from Dauphin Island, the only barrier island located offshore of coastal Alabama. The island protects mainland Alabama's coastal communities and resources from storms, while also providing recreational opportunities and other economic benefits to the local community and the state. Extreme events have severely impacted the island over the past decades, including most recently Hurricanes Ivan, Katrina, and Isaac, as well as the *Deepwater Horizon* oil spill. To combat beach erosion, hard structural solutions have been implemented, including bulkheads and seawalls (on the western end of the island) and rock revetments and bulkheads (eastern end) (Morton et

FIGURE Multiple structures (a seawall, revetment, and failed groin field) were unable to stop erosion at Longboat Key, Florida. These structures were later removed and the beach was nourished. Ultimately, permeable groins were added and the nourishment program continued. SOURCE: Morton et al., 2004.

al., 2014). Recently constructed seawalls to protect homes on the western end of the island (some of which have been rebuilt multiple times) are believed to be worsening erosion downdrift and to the west of the homes (Janasie and Deal, 2015). Coastal development and settlement patterns with their varying needs (i.e., permanent residents on the eastern end and vacationers and renters on the western end) have compounded the physical differences in the morphological evolution and stability of the eastern and western parts of the island. Nonetheless, residents have an extremely strong sense of place and a desire to preserve their way of life, including hopes to develop additional waterfront infrastructure and support services (Janasie and Deal, 2015). With continuing chronic erosion, an average elevation of only 7.2 feet, and a width of barely a mile at its widest point, Dauphin Island is highly vulnerable to changing conditions, such as accelerating sea level rise and predictions for increased frequency of intense hurricanes. A collaborative study among the U.S. Geological Survey, USACE, and the State of Alabama (funded by the National Fish and Wildlife Foundation) is currently under way to investigate options for increasing Dauphin Island's resiliency (USGS, 2017).

BOX 2.4
Recent Examples of Coupled Modeling of the Natural-Human System in Several Coastal Regions

U.S. East Coast: Coupled natural-human system models of the coastal zone have recently applied concepts from nonlinear dynamics and bioeconomics to analyze erosion and storm damage (McNamara and Werner, 2008a; Smith et al., 2009; Gopalakrishnan et al., 2016). One example of this work addresses the decision of a community to nourish its beach. Wider beaches provide recreational amenities and convey storm-protection benefits on homeowners, creating an economic incentive to rebuild eroded beaches. The width of the beach is a function of natural erosion dynamics and the engineering intervention to rebuild the beach with sand dredged from another location. The beach width and property values thus evolve together in a coupled natural-human system (Smith et al., 2009). Current empirical parameterizations are largely based on barrier-island systems in the U.S. Mid-Atlantic region (McNamara and Werner, 2008a; Gopalakrishnan et al., 2011; McNamara et al., 2015).

U.S. West Coast: Another example involves community decision making along the U.S. West Coast, where sea level rise, increased storminess, and recent development have left communities more susceptible to flooding and erosion. The work involves sustained engagement with the Tillamook County Coastal Knowledge-to-Action Network[a] that comprises researchers, students, outreach specialists, and coastal community members in Tillamook County, Oregon. The Knowledge-to-Action Network is co-developing a scenario-modeling tool to explore adaptation strategies for reducing vulnerability to coastal hazards. Probabilistic simulations of extreme total water levels, long-term coastal change, and storm-induced erosion are considered, and a range of alternative futures are imagined related to policy decisions using input from stakeholders. The impact of both policy scenario narratives and climate change scenarios on a range of stakeholder-defined metrics (e.g., property values, or desire for beach access) can then be assessed (Lipiec et al., 2018).

U.S. Gulf Coast: Key interactions in the coupled natural-human system have also recently been applied to project plausible future change in Louisiana. The Coastal Louisiana Risk Assessment (CLARA) model, for example, calculates statistical estimates of the likelihood of property and infrastructure damage based on a set of high-resolution hydrodynamic storm surge and wave inputs, which in turn build on simulated changes to the coastal landscape, wetlands, and vegetation (Fischbach et al., 2012; Johnson et al., 2013). In addition to considering different plausible scenarios of sea level rise, land subsidence, and other environmental drivers looking 50 years into the future, the CLARA model also incorporates uncertainty related to the performance of hurricane protection systems (e.g., levee fragility), as well as a range of population projections for the Louisiana coast (Fischbach et al., 2017).

[a] More information is available at: http://envision.bioe.orst.edu/StudyAreas/Tillamook.

tem of models, however, does not yet consider dynamic feedbacks between the human and natural systems, except for the effects of specific restoration or risk-reduction projects considered for the master plan.

Coupled models can be perceived as oversimplifications of human agency and the human decision environment. The modeling approach, however, does not require distilling human decision making into a single deterministic driver. It can account, for example, for heterogeneous preferences and cultural influences (Smith et al., 2010), both optimizing and heuristic decision rules (Slott et al., 2008; Smith et al., 2009), heterogeneous information and beliefs (McNamara and Keeler, 2013), voter participation in political processes (Mullin et al., in press), and stochasticity (McNamara et al., 2015).

Although such models are still in their infancy, these tools provide a means for beginning to consider a range of different future scenarios. Some examples from various regions of the United States are listed in Box 2.4.

CHAPTER 3

Research Gaps

The dynamic nature of the coastal environment and predominance of coastal development and infrastructure located along the Gulf Coast at low elevations—in association with estuaries, river channels, floodplains, wetlands, barrier islands, and inlets, all of which are responding to rapidly changing conditions—puts the natural (physical and ecological) and human components of the complex coastal system at odds in ways that significantly alter the future evolution of both the components and the system as a whole. For example, when the natural system begins to change in ways that are undesirable to people, decisions are often made to modify it to prevent change and protect valuable infrastructure. This further alters how the natural system evolves. In this way, the natural and human components are tightly coupled, influencing each other through time. Although the two systems can be considered in isolation, a true understanding of how each will evolve in the face of changing environmental conditions necessitates an understanding of the ways in which each system feeds back into the other. Understanding these feedbacks becomes especially challenging when environmental conditions change at unprecedented rates, as is projected to occur for the Gulf Coast (see Chapter 2 for additional discussion). When this occurs, there is the potential for both the natural and the human systems to evolve in ways that cannot be predicted by past empirical data, leading to the need for novel approaches to project and understand future changes in the Gulf Coast's coupled coastal system.

Understanding long-term coastal zone dynamics under changing conditions within the coupled natural-human coastal system will require addressing research gaps related to (1) the physical and ecological components of the natural system (as well as interactions and feedbacks among these components), (2) the human system, and (3) the dynamics of interactions and feedbacks between the natural and the human systems. In this chapter, high-priority research gaps are presented that, if addressed, would transform the ability to understand the Gulf Coast coupled system and the capability to project its future evolution.

Given the interconnectedness of the coupled coastal system, all of the research gaps that were identified require some level of consideration of the entire system. As with Chapter 2, this chapter begins with the holistic view. Subsequently, research gaps that involve one-way interactions between various aspects of the system are grouped by considering the disciplinary lens through which they can best be addressed. Research gaps that involve the consideration of feedbacks among all system components are discussed separately in the last section. Each research gap presented is accompanied by a selected list of associated research questions that, if answered, would address the gap.

It is important to note that the research gaps are not provided in any type of ranking, nor are the associated research questions listed with them. The research gaps, and to some extent the number of associated research questions, do tend to reflect the fact that there is a much greater amount of scientific research that has already been conducted on the physical and ecological components; the research on the human system is less well-developed. This leads to varying levels of granularity in the questions related to different research gaps.

THE COUPLED NATURAL-HUMAN SYSTEM

Physical drivers (e.g., sea level rise, episodic storms) can cause not only physical modifications such as coastal erosion and landform migration, but also ecological alterations such as wetland loss and displacement of biological communities. Such changes can trigger human responses that, in turn, lead to further adjustments in the physical and ecological systems. Human activities can also be drivers for change in the natural system. For instance, the impacts of coastal development on coastal ecosystem function or physical processes can generate feedbacks. Currently, understanding of these feedbacks for the coupled natural-human system in the Gulf Coast is limited.

Projecting the range of potential future behaviors of the coupled natural-human coastal system is challenging, particularly for longer timescales such as decades and centuries. Systems models can be applied to consider a range of scenarios representing a future without action or with a range of proposed coastal restoration and risk-reduction projects implemented, yielding future forecasts or projections of system behavior. Developing such projections depends not only on sufficient understanding of the physical, ecological, and human processes involved, but also on an understanding of the interactions and feedbacks among these components and accurate projections of external factors that could influence the system. Models that couple the natural and the human components of the system to explore its holistic evolution are under development, though such efforts are still in their infancy, and the encoded relationships are at times simplistic. Furthermore, such complex and nonlinear models can also give rise to emergent (even chaotic) behavior and may need to be constrained with observations to lead to usable results. Although the performance of simulation models over the historical long term can be evaluated using hindcast analysis, verification of future projections is exceedingly difficult (and rarely undertaken, though potentially possible). Evaluation requires long-term iteration between making future projections and subsequently collecting targeted long-term datasets for comparison over long timescales.

THE NATURAL SYSTEM

Natural coastal and nearshore systems have been the focus of several white papers that lay out future research agendas (e.g., Elko et al., 2014). This section refines the information from those white papers, focusing on research gaps that are most pressing to further understanding of the Gulf Coast. The importance of studying the natural and the human systems in tandem, and their feedbacks and interactions, is also emphasized.

Physical Processes

Physical processes that drive changes in the natural coastal system occur over varying timescales and range from episodic (e.g., hurricanes, local or distant rainfall events) to longer-term processes (e.g., rising sea level, subsidence). These processes also act across a range of spatial scales and cause short-term disturbances such as flooding, shoreline erosion, and changes in water quality, as well as longer-term changes in the landscape such as river channel migration, wetland loss, and the migration of barrier islands and tidal inlets.

Sea Level Rise

Future sea level rise along the Gulf Coast cannot be viewed in isolation from global sea level rise, even though there is an important role for regional considerations. Understanding the oceanic component of sea level rise along the Gulf Coast is foundational to projecting the future evolution of the coupled natural-human coastal system. Beyond global-scale influences, sea level rise will depend on regional changes in temperature and salinity, dynamic changes associated with ocean circulation, and subsidence (which is particularly important in the northern and western Gulf Coast and discussed in the next section). While there are numerous tide gauges along the entire Gulf Coast, the majority have produced time series that are incomplete or too short (NOAA, 2018), probably reflecting the fact that many of these instruments were never established to monitor relative sea level change. Consequently, there is only a sparse network of tide gauges that have operated long enough (i.e., more than 30 to 50 years; Pugh, 1987; Douglas, 1991; Shennan and Woodworth, 1992) to track rates of relative sea level rise in a meaningful way.

Perhaps the greatest challenge to monitoring sea level rise in coastal zones is that current space-based sensors are unable to produce meaningful signals of sea level change in the near-coastal ocean from satellite altimetry (NASA, 2016). The Surface Water and Ocean Topography[1] (SWOT) satellite mission, scheduled to be launched in April 2021, may offer a new means to monitor climate-driven sea level rise near the coastline.

Research Gap 1: Current datasets, monitoring systems, and approaches are insufficient to track and understand how the oceanic component of sea level (i.e., excluding subsidence) is changing along the Gulf Coast and to predict how it will change in the future.

- What are the spatial and temporal variations and impacts of steric sea level rise (i.e., due to changes in ocean temperature and salinity, including the effect of changes in freshwater discharge from large rivers)?
- What are the spatial and temporal variations and impacts of dynamic sea level rise due to factors such as shifts in the position of major ocean currents (such as the Loop Current) and shifts in winds that force water on and offshore?
- What are the spatial and temporal variations and impacts of changes to the Earth's gravitational field, primarily due to melting ice sheets?

Subsidence

Subsidence rates are higher than the rate of climate-driven sea level rise along considerable portions of the Gulf Coast, notably its western half (e.g., Nienhuis et al., 2017). Although this will likely change in the future because global sea level rise rates are projected to accelerate, subsidence will remain a compounding factor for many regions. There are compelling scientific and practical reasons to reduce the uncertainty in the quantification of subsidence rates and their spatial patterns. While considerable progress has been made over the past decade in understanding subsidence mechanisms and rates, substantial gaps in such understanding still remain.

The rod surface-elevation table-marker horizon (RSET-MH) method (Webb et al., 2013) has led to major advances in understanding shallow subsidence rates in coastal Louisiana (e.g., Jankowski et al., 2017; Osland et al., 2017). This method could be expanded to encompass wetlands along the entire Gulf Coast, similar to the broad and comprehensive Coastwide Reference Monitoring System network in Louisiana.[2] Although there is a considerable network of global positioning system (GPS) stations along the Gulf Coast (Karegar et al., 2015; Yu and Wang, 2016), the network has inadequate spatial resolution to resolve localized phenomena such as faulting and fluid extraction. It is critical that newly

[1] More information about SWOT is available at: https://swot.jpl.nasa.gov.

[2] More details about the Coastwide Reference Monitoring System are available at: https://www.lacoast.gov/crms2/home.aspx.

installed GPS stations have well-documented anchor depths so that the interpretation of their records is straightforward. To date, this has not always been the case.

One of the obstacles toward progress in subsidence research has been the fact that this problem has traditionally been investigated in a monodisciplinary fashion. Recently, concerted efforts have been undertaken to integrate different measurement techniques in a systematic way through experiments known as "subsidence superstations" (Allison et al., 2016). Superstations can integrate measurements obtained with different techniques, such as GPS, optical fiber strainmeters, and RSET-MH, with stratigraphic and geotechnical analyses of continuous sediment cores. These new integrated data are expected to lead to a host of new insights.

There are also opportunities to introduce other new measurement techniques, including corner reflectors installed in wetland environments to be used by satellites for InSAR (Interferometric Satellite Aperture Radar) measurements, a space geodetic technique that can track changes in surface elevation. This method has been used successfully elsewhere (Strozzi et al., 2013), with emerging utilities around the Gulf Coast (e.g., Jones and Blom, 2014; Jones et al., 2016). However, while traditional InSAR has been applied successfully in urbanized settings (Dixon et al., 2006; Jones et al., 2016), extending it to wetlands has proven challenging due to the scarcity of hard surfaces suitable for signal reflection (though it has been done successfully; e.g., Jones and Blom, 2014).

Significant gaps in understanding remain with regard to the role of faulting and fluid extraction as contributors to subsidence along the Gulf Coast. While progress can be expected from recent studies of fault patterns based on industry-grade 3D seismic data, little is known about the contribution of hydrocarbon extraction to subsidence in areas with significant production (Morton and Bernier, 2010; Kolker et al., 2011; Chang et al., 2014).

There is a need for dedicated studies that link the production history of oil and gas fields to subsidence rates and the impacts of shallow fluid withdrawal, both to understand drivers of historic change and to inform future projections. In addition, because shallow subsidence due to compaction of Holocene sediments is a dominant process in many portions of the Gulf Coast (notably in coastal Louisiana; Törnqvist et al., 2008), there is a particularly great need for geotechnical models that include loading, dewatering, and other compaction processes in the prediction of subsidence rates. To date, there have been few studies of this kind, so there is significant scope for continued work, especially for studies that use newly collected data to calibrate and validate predictive geotechnical models.

Integrated coastal subsidence modeling is currently still in its infancy (Allison et al., 2016), and existing models typically focus on just one of the many processes previously discussed. Some modeling efforts, such as glacial isostatic adjustment modeling, have seen considerable progress (Love et al., 2016). An example of an underexplored yet critical area is geotechnical modeling of shallow, highly unconsolidated coastal deposits. Ultimately, subsidence models developed by one specific community will need to become fully accessible to other communities (e.g., geophysics versus geotechnical engineering).

Research Gap 2: The causes, rates, and patterns of subsidence along the Gulf Coast are not sufficiently well understood to allow for accurate prediction at the local to regional scale.

- How can next-generation subsidence modeling approaches that integrate models from different fields improve the capability to accurately predict subsidence?
- What are the contributions of faulting and fluid extraction to subsidence, as revealed through studies of oil and gas field production histories and denser networks of instruments (e.g., GPS) that monitor deeper crustal processes?
- How can geotechnical models contribute to a better understanding and prediction of shallow subsidence processes?

Episodic Coastal and Riverine Flooding

Flooding events along the Gulf Coast are related to the effects of major landfalling hurricanes and other storms, but are also affected by precipitation and related changes in river discharge. Storms can cause major devastation and their flood damage potential will increase in the future due to concomitant sea level rise. Studies have suggested that sea level rise, more than changes in storm climatology, will be a dominant driver of future hurricane-induced flooding (e.g., Woodruff et al., 2013). However, successfully predicting how sea level rise will affect storm-induced flooding along the Gulf Coast will require improved understanding of the processes involved, particularly in areas farther inland where precipitation and related changes in river discharge are also important.

Climate change stressors (e.g., changes in precipitation intensity and patterns, sea level rise, potential backwater effects) are likely to change the hydrology of watersheds draining into the Gulf Coast. Humans also alter the hydrology of the system by diverting freshwater for uses such as drinking water or flood control, modifying land use patterns (e.g., development, agriculture, industry, deforestation, channelization), constructing and operating dams, and developing river diversions that direct water away from large rivers and into coastal embayments for coastal land creation. Changes in watershed hydrology alter sediment availability, salinity, and nutrient flow (including those from upstream runoff) into the coastal system, which contribute to both physical and ecological alterations (e.g., shoreline erosion, subsidence, wetlands accretion, changes in river channel morphodynamics, barrier island formation, oyster reef formation, and estuarine and marine eutrophication). Bio-physical changes, in turn, influence important ecosystem services such as storm damage reduction, flood control, waterfowl, and fisheries. Many major changes to the coastal system are directly or indirectly linked to inland water resources (NRC, 1994).

The effects of human activities on the coast, including urbanization, building of infrastructure, and loss of natural systems such as wetlands, may exacerbate the impacts of storm-driven flooding. However, these effects are not well understood. An emerging area of concern is how the development of levees, floodwalls, storm gates, and other infrastructure designed to reduce flooding impacts in designated areas might impact flooding in other neighboring areas. For example, building storm surge levees may have the effect of redirecting surge and waves from the design area toward neighboring areas, which may then experience adverse effects. Proposals to build a storm surge barrier across the mouth of Lake Pontchartrain show that the barrier could redirect floodwaters toward communities outside the lake, including St. Bernard Parish in Louisiana and Hancock County in Mississippi (Fischbach et al., 2017). In addition, in several locations in Louisiana, efforts to control storm surge through the use of storm surge barriers or gates could exacerbate vulnerability to river flooding—closing the gates to keep coastal flooding out could lead to flooding from the river instead, as the freshwater piles up and has nowhere to go.

Research Gap 3: The combined effects of freshwater input from Gulf Coast watersheds, storm surge, sea level rise, and development on coastal flood hazards are not well understood, thereby limiting the capacity to include and model these effects in predictions of Gulf Coast dynamics.

- How is storm surge affected as the coastal landscape (e.g., landforms, ecosystems, land use) changes in response to sea level rise and coastal development?
- What controls the relative importance of storm climatology and sea level rise in driving the storm flood hazard, including consideration of flooding due to the combined effects of precipitation (and associated river discharge) and storm surge?
- What processes (e.g., astronomical tides, cold fronts, meteotsunamis, coastal evolution, landcover changes associated with ecosystem change) most influence currently observed and future increases in nuisance flooding as sea level rises?
- How do human modifications of the coastline, including urbanization, built infrastructure, gray and green coastal engineering approaches, and land use, affect storm surge and nuisance flood hazards?

Riverine Sediment Transport

Many low-elevation coastal zones worldwide experience reduced sediment input from their drainage basins, commonly due to damming upstream (Syvitski et al., 2005). This is the case in several Gulf Coast rivers (e.g., Rio Grande River) that have seen dramatic reductions in their sediment delivery to the coast (Milliken et al., 2017). Reduced sediment fluxes have been proposed as a major cause of coastal degradation along the Gulf Coast (e.g., Blum and Roberts, 2009) and may be compounded by the embankment of major rivers for flood-reduction purposes. Levees prevent the natural dispersion of sediment across the coastal landscape, resulting in elevation deficits and, in many cases, eventual wetland loss. This problem is particularly acute in the Mississippi River Delta, where planning is currently well under way for mitigation measures via managed sediment diversions at strategically selected locations (LACPRA, 2017). Decreased sediment delivery by rivers also affects smaller bayhead deltas that occupy estuaries (e.g., in Texas) and barrier islands (Anderson, 2007). However, this is less of an issue for Florida's Gulf Coast, which has historically had lower sediment inputs due to relatively small, shallow gradient watersheds (Kumpf et al., 1999).

One poorly resolved issue is whether the present-day sediment flux of the heavily managed Mississippi River, by far the largest sediment source in the Gulf of Mexico, is significantly smaller than in its natural state without human modifications. Many studies suggest this is the case, based on instrumental records that show a distinct reduction of sediment flux during the 1950s when a series of major dams were constructed on the Missouri River, the largest sediment source within the Mississippi River drainage basin (Meade and Moody, 2010). It is possible, however, that historic sediment fluxes were significantly higher than prehistoric fluxes, due to clearing of the drainage basin for agriculture and the associated increase in erosion rates (Keown et al., 1986; Tweel and Turner, 2012b). Resolving this uncertainty would contribute to a better understanding of sediment management within this large region, because it would help clarify the degree to which current landform change is a response to historic changes in sediment supply, current changes in hydrodynamics, or relative sea level rise.

Another poorly resolved issue is the amount of sediment that is available for ecosystem creation and land-building projects across the Gulf Coast, whether "new" sediment recently transported to the Gulf of Mexico via rivers or "existing" sediment from river beds, the seafloor, submarine shoals and tidal deltas, and other coastal sedimentary features. While there has been some progress to date to quantify these resources, much remains unknown about the total volume of sediment that exists for ecosystem creation projects, its quality (e.g., particle size, contaminant potential), and the location of sedimentary resources in relation to coastal restoration areas where sediment is needed.

Along the eastern Gulf Coast, questions also remain about the availability of sediment necessary to sustain coastal systems in the face of sea level rise. These stem from a need to better understand alongshore sediment transport pathways and the impacts of structures (e.g., groins, jetties) on sediment transport pathways, as well as the capacity for biogenic sediment production to supply meaningful amounts of sediment.

Research Gap 4: The relative contributions of naturally occurring and artificially managed riverine sediment delivery (e.g., availability and fluxes), diversion and management activities, and how they impact the evolution of coastal landforms (e.g., river deltas, barrier islands) and ecosystems (e.g., wetlands) is poorly understood.

- How does the present-day sediment flux of the Mississippi River and other Gulf Coast rivers compare with historical and prehistorical fluxes?
- What sources of sediment are available for restoration efforts and in what quantities?
- How will the availability of sediment to wetlands and barrier islands change in the future, and what will be the likely effect of these changes on coastal evolution?

Coastal Sediment Transport and Hydrodynamics

The fate, delivery, and transport of sediments exert major control on the evolution of the coastal landscape. Significant progress has been made in understanding the dynamics of sediment transport on sand and cobble beaches and barrier islands along unprotected coasts. However, understanding of sediment transport remains limited in areas where cohesive sediments or mixed grain sizes are dominant, and in the presence of coastal development and hazard mitigation infrastructure. Formulations of sediment transport are an important component of many coastal evolution models. Currently, sediment transport formulations in wide use within coastal evolution models are empirically based and do not fully represent all governing physical processes. Though there have been significant recent advances in understanding transport at the sand grain scale (e.g., Simeonov and Calantoni, 2011), there is currently an inability to upscale these grain-scale processes to the spatial and temporal scales required for predicting long-term coastal change while retaining the fundamental physics of sediment transport. Recent advances use statistical approaches as a way to move toward addressing this issue (Palmsten et al., 2017) and have the potential to provide an important additional means for predicting patterns of coastal behavior at large spatial and temporal scales under changing conditions. It is also important to better understand the critical role of biota (e.g., microphytobenthos, dune grasses, mangroves) in influencing sediment transport and to incorporate improved understanding of this role into sediment transport formulations.

Developing reliable projections of long-term coastal evolution will also require improvements in the ability to project future hydrodynamic conditions. The recent development of a novel statistical approach that uses climate-model-derived wind field output to predict future wave climate (e.g., Camus et al., 2014a,b) represents an opportunity to improve the representation of such processes in future projections (Antolínez et al., 2016). It is critical to develop more such methods for making reasonable projections of hydrodynamic conditions. In addition to wind and wave forcing, regional-scale hydrodynamics such as the Loop Current have the potential to influence large-scale evolution of coastal features, yet the understanding of hydrodynamics at this scale and the relationship among such processes is insufficient to include their effects in modeling efforts.

Research Gap 5: Limited understanding of sediment transport processes and uncertainties in predicting future hydrodynamic conditions hampers the ability to project long-term coastal evolution.

- What improvements can be made in methodologies to predict sediment transport in the coastal zone?
- How do biota influence sediment transport and what modifications to sediment transport formulations are necessary to include these effects?
- What are the main drivers and sensitivities connecting regional-scale hydrodynamics to large-scale coastal evolution?
- How will hydrodynamic conditions (e.g., waves, tides, inundation, currents) change in the future?
- How might shelf-scale oceanographic processes control coastal evolution (e.g., effects of the Loop Current)?

Evolution of Coastal Landforms and Embayments

Coastal landforms and embayments in the Gulf (e.g., mainland shores, deltaic plains, chenier plains, river deltas, barrier islands, peninsulas, tidal inlets, bays, and lagoons) change and evolve over a range of timescales in response to waves, tides, and currents that drive sediment transport (both in fair weather and during storms), as well as sea level rise, subsidence, ecological processes, and human processes. As sea level rise accelerates and environmental conditions such as storminess, precipitation, and temperature shift in the future, the dynamic coastal landforms and embayments of the Gulf Coast will respond by evolving even more rapidly than they have in the recent past.

The ability to simulate and forecast the long-term evolution of coastal landforms and embayments is critical to understanding the coupled natural-human system and would draw directly from progress toward addressing Research Gaps 1 to 5 (described in previous sections). Models that forecast evolution of the physical coastal system have progressed over the past decade, though the capability to make regional-scale, long-term predictions of coastal change across landscapes that include a range of different types of coastal environments is still far away. Development and testing of these models also hinges on the availability of recent, synoptic-scale bathymetry and nearshore topography for the Gulf Coast.

Contributions of sediment from rivers, alongshore sediment transport, and shoreface erosion, along with human activities that modify them, determine the rates at which sediment is supplied to or lost from the coast across a range of timescales. Assessing these contributions is essential for modeling and projecting coastal evolution; however, these contributions are not well understood (see Research Gaps 4 and 5). The amount of sediment supplied to wetlands and barrier islands is especially critical in determining whether they can accrete and maintain their position and function as sea level rises. Sediment reduction or blocking can have long-term consequences (e.g., Rogers et al., 2015). Climate change–induced shifts in dominant vegetation type (e.g., Zarnetske et al., 2012; Armitage, 2015) or characteristics (e.g., Emery et al., 2015) on barrier islands and mainland, or estuarine shores, can alter patterns of sediment transport and deposition (e.g., Hacker et al., 2012; Durán and Moore, 2013). This can lead to landscape alterations that affect ecosystem function and coastal vulnerability. Furthermore, although it is well known that vegetation can act as a strong stabilizing agent against erosion in a variety of ecosystems (e.g., forest, riparian, agricultural), much remains to be understood regarding the specific interaction of barrier island vegetation and substrate accretion and stability (e.g., Maun, 1998; Tsoar, 2005). Wave climate changes can also alter sediment transport, affecting patterns of shoreline erosion and accretion and leading to changes in coastline shape (e.g., Slott et al., 2006; Moore et al., 2013; Thomas et al., 2016), and current and future coastal vulnerability (e.g., Wahl and Plant, 2015). In addition, understanding the role of preexisting geologic conditions (e.g., Rodriguez et al., 2004; Mallinson et al., 2010) and feedbacks between physical processes and ecological processes (e.g., Short and Hesp, 1982; Psuty, 2008; Houser and Hamilton, 2009) along the Gulf Coast is still evolving. These all lead to a need to identify coastal tipping points, such as the loss of wetlands and barrier islands, in light of anticipated rapid changes in environmental conditions in the future.

As discussed in Chapter 2, there have recently been important advances in the development of coastal area models. These predictive models require significant computational resources (e.g., Delft3D[3] [Lesser et al., 2004], XBeach [Roelvink et al., 2009]) and typically are only capable of simulating short-term coastal change. Various schemes, however, have been proposed to reduce computational time (e.g., time-averaging the equations over timescales of tides or longer to allow for longer model time steps or morphological factors that accelerate bathymetric change) to arrive at models that can be executed rapidly for very long periods of time (Roelvink et al., 2006). Further reductions in the computational needs of predictive coastal area models are also possible (e.g., new computational methods and future computational power) and would enhance their utility by allowing simulation of coastal evolution across longer time and spatial scales.

[3] Additional information about Delft3D can be found at: https://oss.deltares.nl/web/delft3d/about.

In addition to developments in predictive coastal area models, there have been advances in reduced-complexity modeling approaches (e.g., French et al., 2016) that include explicit treatment of key processes (e.g., Coastline Evolution Model [CEM; Ashton and Murray, 2006]; COastal Vector Evolution Model [COVE; Hurst et al., 2015]; ShorelineS [Roelvink, personal communication[4]]). These models are computationally efficient and lend themselves to being coupled with other system models to make predictions over large spatial and temporal scales. These reduced-complexity models, however, can involve significant simplifying assumptions and may need extensive, site-specific calibration. Further development of these types of models is needed, for example through data assimilation approaches (e.g., Vitousek et al., 2017), to expand the temporal and spatial scales across which they can make realistic predictions.

Finally, opportunities exist to combine the strengths of coastal area, coastal evolution (planform), and landform evolution models (e.g., of barrier islands, wetlands, estuaries) by coupling them in appropriate ways. For example, importing relevant information from high-resolution hydrodynamics-based coastal area models into the input conditions of other types of models can more effectively address long-term evolution of the coastal system (e.g., van Maanen et al., 2016). In addition to coupling different types of coastal models, it is important to include feedbacks between physical processes and ecological processes in models of coastal evolution. There has been some progress toward this end, such as the inclusion of connected barrier-marsh and bay processes in landform evolution models (e.g., Walters et al., 2014) and inclusion of the effects of vegetation on bed friction in XBeach (Passeri et al., 2018).

Ensemble approaches, including stochastic forecasting such as Monte Carlo methods, have been used effectively in other branches of geoscience and engineering, as well as in meteorology, to characterize uncertainty associated with variable or unknown processes, or to consider the range of predictions derived from the use of different modeling approaches. Many studies of long-term coastal change already adopt an ensemble approach to address a lack of constraints on the exact sequencing and spatial distribution of physical, ecological, and human drivers (e.g., Barkwith et al., 2014). Extending the ensemble approach to include consideration of the range of future behavior predicted for the same future conditions by different long-term models and modeling approaches, such as those so commonly used in climate modeling and hurricane forecasting, would provide a more comprehensive means for developing and assessing the reliability and range of future predictions (e.g., Kirwan et al., 2010). Pairing this approach with long-term monitoring of landscape change would allow comparisons with and among model results over time, facilitating model development and improvements in capability.

[4] Communication during committee's open session in New Orleans, Louisiana, on September 18, 2017, with Dano Roelvink, IHE Delft Institute for Water Education.

Data assimilative approaches have also been explored in related geoscience fields to constrain and guide model solutions or determine poorly known boundary conditions or parameter values (e.g., surf zone bathymetry estimation [Wilson et al., 2014], sediment distribution estimation [Wiberg et al., 2015]). Such approaches can lead to model configurations that are better suited to accurate predictions. They can also result in an understanding of crucial observations and can therefore inform the design of long-term observational programs.

Research Gap 6: There is a critical need to understand and project the future response of coastal landforms and embayments to changing climate and the conditions under which they will no longer be able to keep pace with relative sea level rise.

- What are the critical tipping points beyond which barrier islands, river deltas, and wetlands will be unable to keep up with relative sea level rise, and what indicators can be used to reveal when the system is approaching them?
- How will sediment budgets, transport rates, and patterns change in the future and what will the impacts be on the evolution of barrier islands, mainland shores, and wetlands?
- How do feedbacks between biota and sediment dynamics, under current and future conditions, affect the structure and evolution of coastal landforms along the Gulf Coast?
- How has the Gulf coastline responded to changes in wave climate, rates of relative sea level rise, and sediment supply in the past, and what effects might future changes in these variables have on coastal evolution?
- How can future evolution of the physical coastal system (e.g., open-ocean coastline, estuaries, barrier islands, river deltas) be reliably modeled and projected over long timescales and large spatial scales?
- What monitoring and observational studies can be developed to capture and understand changes as they begin to happen even more rapidly in the future?
- How can existing approaches to modeling the physical system be improved or new approaches developed (e.g., by incorporating data assimilation approaches and the expanded use of ensemble modeling) to encompass a full range of future landform and embayment projections?
- What improvements in the reliability of modeling future landform and embayment projections can be realized through iteratively incorporating new understanding of the physical system?
- What are the best means for testing and further improving the reliability of modeled future projections of landforms and embayments?

Ecological Processes

Gulf Coast ecosystems continue to evolve over a decadal to centennial scale. These changes are inherently linked to changes in the physical system and are often human induced, arising from factors such as changes in the built environment, industrial practices along the coast, and climate change. Placing the ecological component within the coupled natural-human Gulf Coast system first entails understanding how ecosystems function under natural conditions, and then how human alterations affect ecosystem function.

EFFECTS OF CURRENT AND FUTURE PHYSICAL FORCING AND ENVIRONMENTAL CONDITIONS ON COASTAL ECOSYSTEMS

The Gulf Coast is characterized by sharp longitudinal contrasts in terms of sediment, nutrient and freshwater inputs, and resulting environmental conditions for its biota. Although these natural environmental gradients strongly affect the composition, structure, and species richness of coastal ecosystems, the impacts of these gradients on ecosystem function and resilience are not well understood. For instance, understanding the functional changes from the phytoplankton-dominated western Gulf to the macrophyte-dominated eastern Gulf (Anton et al., 2011; McDonald et al., 2016b) is far from complete. While the knowledge of compositional and structural shifts in Gulf Coast ecosystems is detailed, it is not well known how these shifts translate into functional changes.

Along the Gulf Coast, physical forcing varies over diverse timescales (e.g., daily, weekly, monthly, seasonal, inter-annual). Episodic events, such as hurricanes and cold fronts, can also induce substantial temporal variability. However, the impacts of physical temporal variability on the dynamics and function of coastal ecosystems are not fully understood. For instance, more research is needed on how Gulf Coast ecosystems respond to rising water levels driven by global sea level rise and subsidence, which takes place over years to decades; changes in wind-driven setup, which take place over hours to days (but also have a multi-decadal component); and river floods, which take place over days to months; and the interactions among these.

Coastal shorelines in the Gulf of Mexico will continue to experience impacts from coastal development such as residential, commercial, and energy infrastructure; legacy impacts associated with this infrastructure; hydrological alterations; and exploitation of natural resources. All of these impacts can cause deterioration and compromise the resiliency and sustainability of coastal ecosystems. There have been meaningful advances in understanding the impact of coastal development on Gulf Coast ecosystems, but much remains to be learned. A warming climate and rising sea level will certainly cause further coastal ecosystem change in the decades ahead. Changes in watershed hydrology will also affect sediment flux, resulting in wetland accretion or erosion and affecting services provided by wetlands and other coastal ecosystems. There is substantial evidence (Osland et al., 2013; Weston, 2014) that anthropogenic impacts on Gulf Coast ecosystems will likely become more pronounced in the future, but how this may alter their dynamics and function is less well known. In particular, the compounding effects of coastal development and climate change on ecosystem dynamics and function are poorly understood, yet are crucial to understand and co-manage ecosystems and human populations along the Gulf Coast in future years. It is also important to put the interactive effects of development and climate change on coastal ecosystems in the context of associated changes in the physical system (e.g., open-ocean coastline, estuaries, barrier islands, river deltas) to acquire a holistic understanding of the future evolution of the natural-human coastal system in the Gulf Coast. This can be achieved through the use of integrated models that synthesize the response by physical and ecological systems to climate change and coastal development.

Research Gap 7: There is limited understanding of the individual and combined effects of current environmental gradients, physical forcing, climate change, and coastal development (including energy-related infrastructure) on Gulf Coast ecosystems.

- What are the effects of current and predicted future (e.g., climate change, coastal development) gradients in physical and environmental conditions (e.g., sea level rise, sediment loading, water clarity, tropicalization, ocean acidification, changes in storm frequency and intensity, watershed hydrology) on the distribution and abundance of key ecosystems along the Gulf Coast, including oyster reefs, salt marshes, submerged grass beds, and barrier islands?
- How are ecosystem services (e.g., fisheries productivity, shoreline stabilization through wave attenuation, filtration and removal of pollutants, mitigation of climate change through carbon sequestration) affected by these current and predicted future gradients?
- How prominent will the effects of relative sea level rise be in relation to other stressors such as coastal development, and what are the interactions among these processes and the consequences for ecosystem survivability?
- What are the most important interactions between and among physical forcing, coastal development, and climate change driving changes in environmental conditions and ecosystem response, both in terms of their structure and function?

Effects of Strategic Natural Resource Conservation and Restoration on Coastal Ecosystems

There is evidence to suggest that strategic natural resource conservation (e.g., functional hotspots) can help attain sustainable and resilient coastal systems (McDonald et al.,

2016a). However, knowledge on how to identify functional hot spots, and their viability and effectiveness to maintain ecosystem function under current and future climate regimes, is still in its infancy. Restoration efforts (e.g., creating land through river diversions) can fill a similar role. These efforts come in many forms and can vary widely in design, cost, execution, evaluation, and performance. Unsuccessful efforts are often grounded in a lack of understanding and consensus about desired recovery targets, as well as limited understanding of the response of the resource to existing environmental conditions (Sparks et al., 2013). Research on the response of restored resources to environmental conditions on the developed coast under current and future climates, as well as criteria for the establishment of restoration benchmarks sought relative to current or past conditions, is needed for more effective outcomes and higher success in restoration efforts.

Research Gap 8: The understanding of strategic natural resource conservation and restoration activities for effective coastal management is limited.

- Which functional hot spots best contribute to the preservation of ecosystem resources, function, and services in developed coasts under current and future climate conditions?
- How do restored resources respond to stressors from human development and climate change?
- What are reasonable, realistic benchmarks for gauging restoration success in developed coasts under current and future climate conditions?

THE HUMAN SYSTEM

Understanding the evolution of the coupled natural-human coastal system necessitates an understanding of the components of the human system that interact with and feed back on the natural (physical and ecological) system. The examples of human dynamics and decision making discussed in Chapter 2 underscore the lack of scientific basis to predict significant human processes with any degree of confidence over decadal to centennial timescales. These types of projections will be particularly challenged if thresholds and tipping points are exceeded, leading to new equilibrium states for any of the coupled natural or human system components. An argument can be made that many of the most important human changes in the 20th century are unique to that time, and such significant changes are not likely to occur again. It may be reasonable to model human processes with the expectation that fundamental ways of living over the next 10 to 200 years will be similar to today. However, many environmental changes in the coastal zone will be unprecedented and may lead to entirely new ways of living in the coastal zone and new patterns of economic activity—these will need to be considered in modeling efforts. The essential features of the human system that models will need to consider can be grouped into three broad areas: decision making and adaptation, the built environment, and human migration.

Decision Making and Adaptation

Individual households and governments sometimes invest in durable goods to adapt to climate change. Households might purchase high-clearance cars or trucks to transit areas of minor flooding, replace carpet with tile to deal with nuisance flooding, or raise homes on pilings to avoid damages from storm surge. Governments might build a seawall or fund a beach nourishment project to mitigate against storm risk or erosion associated with sea level rise. Future investments in durable goods may come in the form of new innovations or adaptation technologies based on ideas that are not yet known or cannot yet be feasibly implemented. Which durable goods individuals and governments choose, and when they choose them, ultimately influences the long-term evolution of the coupled natural-human system.

Adaptive Decision Making

Most efforts to model human adaptation measures do not incorporate insights on how humans make decisions. Models of defensive adaptation expenditures, such as sea walls, levees, beach nourishment, and other gray, green, or hybrid interventions, typically impose decisions on the physical system. Similarly, to the extent that there is modeling of zoning, setback rules, or building codes, the focus is on modeling the effects on the physical, ecological, or human system. Less understood in the long-term evolution of the coupled natural-human system is how these decisions are determined endogenously (i.e., within the system in response to feedbacks, rather than external to the system). Imposing a particular decision or defensive expenditure on the coupled system without a model of how future decisions will adapt to future conditions will likely not produce complete or useful scenarios to analyze long-term changes in the coastal zone. A handful of studies have coupled human adaptive decisions with physical coastline change models and have focused on beach nourishment along sandy coastlines, including property value and beliefs about climate change as socioeconomic drivers (McNamara and Keeler, 2013; Williams et al., 2013; McNamara et al., 2015), or both policy and climate change scenarios on a range of stakeholder-defined metrics, such as number of affected buildings or considerations of beach access (Lipiec et al., 2018).

Modeling long-term coastal zone change requires understanding a broad set of mechanisms that drive a range of defensive expenditures and development restrictions, as well as an understanding of the feedbacks that these decisions create on the physical and human systems, particularly how individual decisions evolve with associated human-imposed physical change. More research is needed to apply insights to coastline types and adaptation settings beyond those included in previous studies. More data are needed regarding state, federal, and community-level defensive expenditures and decisions to restrict development. While there are some datasets that track federal payouts on disaster relief (FEMA, 2018), flood insurance, and expenditures on projects such as beach nourishment (PSDS, 2018), datasets that compile federal, state, and local decisions about defensive capital expenditures and local development restrictions are sparse and not uniformly tracked over time and space.

Research Gap 9: There is a need to understand how decisions about the built environment will be affected by coastal change and how these decisions create feedbacks between the natural and human systems.

- How do flooding events influence decisions about future infrastructure siting? How does existing infrastructure influence current and future flood hazard?
- How do decisions about coastal development and abandonment alter the natural evolution of coastal landforms, and what are the feedbacks on subsequent behaviors and decisions?
- How will existing policies or policy changes that provide incentives or disincentives to develop or redevelop (e.g., flood insurance, zoning, buyouts) influence future decision making?
- How willing are coastal residents to pay higher taxes to support defensive capital expenditures or other adaptation interventions to support existing coastal communities?
- At the household level, what decision-making process and what circumstances lead to investments in adaptation measures such as structure elevation or flood proofing?

The Built Environment

The built environment refers to the places and spaces created by people. These are the areas where they live, work, and recreate on a day-to-day basis, such as buildings, infrastructure, and parks. The built environment can be affected by coastal change as seen, for example, in the damage to energy and port infrastructure, destroyed homes and commercial buildings, and damaged levees and other structural defenses during and in the aftermath of Hurricane Katrina and other major Gulf Coast hurricanes. Subsequent decisions about the built environment can also affect coastal environmental change.

Effectiveness of Green or Hybrid Infrastructure

One important issue that exemplifies the need for a better understanding of natural-human system feedbacks is the use of gray-green (or hybrid) infrastructure. Hybrid approaches use nature-based approaches (e.g., wetland creation) to enhance the effectiveness of built infrastructure (Sharma et al., 2016a). Besides helping traditional, human-made (gray) features become more efficient, nature-based (green) approaches can also help preserve ecosystem function and services. However, like gray infrastructure, nature-based methods also affect the long-term evolution of the coupled coastal system. However, there are critical gaps in understanding how green infrastructure performs from an engineering perspective (e.g., Cunnif and Schwartz, 2015), as well as how these systems evolve over time, particularly in the face of rapid sea level rise, and how this evolution impacts their performance. An understanding of how to best combine both green and gray infrastructure in cost-effective ways is limited, as is reconciling variations in performance between the two (green infrastructure is inherently variable), overcoming differences in gray and green infrastructure useable life span, and mitigating adverse effects that may stem from a combination of approaches.

Even when engineered with spatial uniformity, within a single season green infrastructure can undergo significant life-cycle changes (e.g., marsh plants senescing in winter)—altering its ability to trap sediment and dampen waves and introducing spatial variation along with temporal evolution. These spatiotemporal changes alter the green infrastructure's actual performance in the short and long term (Koch et al., 2009). Because of these natural discontinuities, green infrastructure intended to mitigate chronic or episodic inundation, wave, and erosion hazards may evolve to be more or to be less effective than intended (Feagin et al., 2010), and in some hazard conditions might become ineffective or even increase the hazard locally (e.g., Irish et al., 2014).

Improving such understanding is essential for ensuring resilient coastal communities in a future of increased human pressure and climate change. A better understanding of how to best combine green and gray infrastructure into hybrid approaches lies directly at the intersection of the natural and human systems in the Gulf Coast. However, more work is needed to evaluate the applicability, versatility, and effectiveness of these approaches.

Energy-Related Infrastructure

Given that a large percentage of the U.S. energy infrastructure lies along the Gulf Coast, sea level rise, subsidence, increased hurricane intensity, and loss of wetlands and other ecosystems could lead to direct impacts on infrastructure, such as damage to equipment from coastal erosion or flooding. Other indirect impacts, such as increased costs from raising vulnerable infrastructure to higher elevations or building future energy projects farther inland, thereby increasing transportation costs, can also occur (CCSP, 2007). In addition, vulnerabilities and risks associated with aging energy infrastructure cascade not just into transportation systems, but also into water infrastructure, ecosystems, agriculture, forestry, communities, and social systems (Wilbanks and Fernandez, 2013). With increased risks of coastal hazard impacts to aging energy infrastructure, social vulnerability for communities co-located with them has also increased (Bernier et al., 2017).

With the expected future impacts of changing environmental forcing on the Gulf Coast's dynamic landscapes, it may become important to prioritize the most critical infrastructure. This is relevant not only for energy-related infrastructure, but also for other critical infrastructure such as water, power, and transportation infrastructure. There is a need for more knowledge regarding which infrastructure would need more protection (e.g., levees, floodwalls), which would require better physical resiliency (e.g., low-elevation buildings), which would need more flexibility (e.g., working operations), and which might need to be relocated or abandoned. Determining which option works for different types of infrastructure necessitates development of models that can couple the significant physical (e.g., flooding) and economic factors (e.g., profitability of the infrastructure). For instance, in the case of infrastructure that has to be abandoned, there needs to be discussion of how abandonment would occur, what type of demobilization would be needed, and where debris would be moved. For infrastructure that is to be relocated, there needs to be discussion of where it would be appropriate to move, and what the community, economic, and environmental impacts would be for both the original and the new location.

As an example, Port Fourchon is the largest deep draft oil and gas port along the Gulf Coast, providing services for the offshore oil and gas industry. Recognizing its vulnerability to storm surge and erosion, energy companies dependent on that infrastructure have previously invested in engineered protection, as well as beach and dune nourishment. In addition, Port Fourchon is conducting feasibility studies of a major expansion that would include deeper draft navigation channels, and beneficial use of the dredged sediment has been proposed to create wetland habitat areas and reduce storm surge impacts (Guidry, 2016, 2018; Duchmann, 2018).

Evolution of the Coupled System in the Presence of Energy Infrastructure

A comprehensive, field-based assessment of impacts to coastal habitats from the construction and operation of oil and gas infrastructure (e.g., pipelines, navigation channels and waterways, onshore facilities), emphasized the importance of a variety of interrelated factors in establishing causal relationships between the infrastructure and the response of the natural system (Wicker et al., 1989; Cranswick, 2001; Nairn et al., 2005; Johnston et al., 2009). Depending on the site physiography, the construction methodology employed, and mitigative measures, the impacts attributable to infrastructure ranged from an absence of any impacts to significant direct and indirect impacts. Uncertainty still exists in accurately quantifying the relative contribution of human activities (such as oil and gas operations) to subsidence, versus other natural mechanisms such as tectonics (although natural mechanisms such as faulting may also be caused by fluid extraction), all of which vary across the Gulf Coast (Kolker et al., 2011).

Limited understanding of individual- and community-level decision making under different scenarios of coastal change hinders the ability to constrain the range of possible future outcomes. To model the long-term evolution of the coupled Gulf Coast natural-human system, relevant scenarios about individual and government decisions to invest in durable goods for climate change adaptation need to be developed. There is also a need to devise and evaluate the performance of novel mitigation strategies and emerging types of hazard mitigation infrastructure that are intended to be adaptive and resilient to change. Similarly, understanding the vulnerability of energy and energy-related infrastructure to relative sea level rise and storms, as well as the consequences of infrastructure failure, is limited. This is also true for understanding the direct and indirect causal relationships between energy-related infrastructure and the coupled response of the natural system. Taken together, these issues point to an overarching challenge to improve understanding of how the Gulf Coast's built environment interacts with the natural system, including residential housing, commercial buildings, transportation and energy infrastructure, port infrastructure, and defensive structures such as seawalls and levees.

Research Gap 10: There is a need for better understanding of how coastal changes affect the built environment and which aspects of the built environment are most vulnerable to coastal changes.

- What strategies for coastal development are most cost-effective when considering future climate change, relative sea level rise, and episodic events?
- How do coastal engineering approaches or development restrictions feed back on continued coastal development and the tax base that supports engineering interventions?
- Under what circumstances might radically different ways to develop the coastal zone emerge, and how might those approaches affect evolution of the coupled natural-human system?
- In the near-decadal timescale, what Gulf Coast energy infrastructure is vulnerable to factors such as sea level rise, subsidence, increased hurricane intensity, wetland and other habitat loss, or age-related failure?
- How will construction and operation of infrastructure associated with coastal development and energy infrastructure (e.g., pipelines, waterways) affect evolution of the natural system across different temporal and spatial scales, and vice versa?

Migration

Although study of human migration has advanced in general terms, less understood is the relative importance of driving factors for migration decision making across different regions, groups of people, and types of coastal change processes. There is also disagreement and a knowledge gap with respect to the relative importance of long-term, gradual changes (e.g., relative sea level rise driving the movement of the community of Isle de Jean Charles [HUD, 2017]) as opposed to episodic events (e.g., households moving from New Orleans to Houston and other cities following Hurricane Katrina or evacuation of coastal Florida residents in response to Hurricane Irma) in influencing migration (Colten et al., 2008; Gutmann and Field, 2010; Hauer, 2017). Complicating the problem further are the compounding effects of economic and social stressors and multiple interactive and simultaneous sources of environmental change (e.g., sea level rise over the long term, increasing frequency of nuisance flooding, the sudden shock of a major storm). Some economic and social stressors are external to the coupled system; for instance, major changes in energy markets or the decline of entire economic sectors. Someone employed in a declining sector might be more vulnerable to environmental change or less able to adapt to change due to socioeconomic status.

Some economic and social stressors, however, are internal to the coupled system; a natural hazard that destroys an individual's property could directly trigger a decision to migrate. Scientists and planners are not yet able to project or predict the long-term push-and-pull factors that influence human migration on the Gulf Coast in response to coastal change. This lack of understanding, particularly with respect to the direct and indirect relationships between environmental change and human decision making, make it challenging to incorporate these factors into models of the future coupled natural-human system. Consequently, current modeling of long-term coastal change and human response does not include migration decisions in response to coastal change and associated feedbacks on the Gulf Coast system.

Despite growing attention to migration in response to climate change, there is limited understanding of the drivers of human relocation and migration (both external and internal) in response to coastal change along the Gulf Coast, as well as how these decisions create feedbacks in the coastal environment. To include individual, household, and community migration decisions in response to coastal change in modeling scenarios for long-term coastal zone planning, significant advances in understanding are needed. Developing a long-term, longitudinal[5] dataset to follow Gulf Coast residents, in-migrants, out-migrants, and geocoded information on coastal changes would significantly improve the ability to

[5] Longitudinal studies collect repeated observational data on cross-sectional units (e.g., individuals, groups of people, cities) over time.

understand complex migration decisions and could also provide information on rebuilding decisions and household-level changes in response to coastal change.

Research Gap 11: There is an incomplete understanding of the vulnerability of different Gulf Coast communities to coastal dynamics, how coastal dynamics trigger migration and relocation decisions, and how these decisions create feedbacks to the natural system.

- How will people respond to unprecedented accelerations in environmental change, and how can this understanding be used to identify potential tipping points of human response (e.g., when might at-risk coastal communities across the Gulf Coast be abandoned)?
- How do migration decisions influence the tax base and the ability of communities to fund local public goods and defensive expenditures in response to coastal change?
- How do community-level investments in defensive capital, social capital, and local public goods influence migration decisions, including decisions to migrate into or out of a community?
- When do migration decisions (in or out) decrease or increase existing social capital, trigger further migration, and potentially reach a tipping point that undermines community viability or leads to complete abandonment?
- Which household characteristics, economic factors, and types of social capital drive household decisions to stay or migrate in response to coastal change, and how do these decisions aggregate up to larger population dynamics?
- When does evacuation in response to an acute storm event become a permanent migration decision?
- When people opt to rebuild rather than migrate after an acute storm, how do these decisions influence patterns in housing and urban development?

FEEDBACKS WITHIN THE COUPLED NATURAL-HUMAN SYSTEM

The research gaps previously discussed primarily involve one-way interactions among the various system components—how one aspect of the system (i.e., physical, ecological, or human) is influenced by or responds to other components. As such, these research gaps can be addressed while being viewed primarily through the lens of a specific component. In contrast, this research gap involves the consideration of the entire coupled system. Addressing this research gap entails an adequate understanding of the individual components, but the emphasis is on the understanding of the feedbacks among the components and the resulting evolution of the entire system, which can be best understood and projected through integrated modeling.

Research Gap 12: Understanding how decisions about the built environment and human migration will affect the coupled natural-human coastal system is limited and can be furthered through integrated modeling.

- What are the feedbacks among the coupled natural-human system, and how can they be incorporated into models?
- When will feedbacks reach tipping points that fundamentally change the coupled system?
- How can models account for household decisions in response to novel environmental changes (e.g., types or magnitude of changes that households have not yet experienced, such as unprecedented levels of sea level rise)?
- What are the impacts of gray, green, and hybrid engineering measures on the long-term evolution of coastal systems?
- How can modeling of the coupled natural-human system in the Gulf Coast incorporate the deeply uncertain possibilities of novel infrastructure or technologies that could radically change the experience of living along the coast?
- How do emerging or novel engineering approaches perform relative to traditional structures such as seawalls, groins, and beach nourishment?
- How does the performance of coastal engineering approaches in reducing flood hazards influence people's attitudes or behaviors toward further coastal development?

CHAPTER 4

Barriers to Effective Communication Between Scientists and Stakeholders

There is a recognized gap between scientists and stakeholders in terms of the use of scientific information to influence decision making and to collaborate on the development of usable knowledge and understanding (Bartunek et al., 2001; Bartunek and Rynes, 2014). Here, "stakeholder" is broadly defined as any person or entity with an interest or stake in the issue at hand (EPA, 2017). Stakeholders can be members of the general public, a community, a company or industry, an organization, a government agency, or a practitioner such as a city planner or emergency manager. In the Gulf Coast, stakeholders live in the region and maintain and contribute to the state of the coast, making decisions on a range of issues from coastal restoration to adaptation planning. Moreover, they are likely to be influenced by adaptation planning and management decisions (Alexander, 2013). While this chapter touches on the perspectives of stakeholders in general (especially in the sections Scientist Perspective and Media and Communication), there is particular focus paid to "practitioner" stakeholders working on coastal issues related to resiliency and planning.

Clear and consistent communication between scientists and stakeholders (particularly at the federal, state, and local levels) is of paramount importance to project success (Byrnes and Berlinghoff, 2012). Despite the recognition that evidence-based decision making improves outcomes (Small and Xian, 2018) and that collaboration and coproduction of scientific information is beneficial to the process (Prokopy et al., 2017), effective communication[1] often remains elusive. Task 3 of the committee's Statement of Task (SOT) (see Box 1.1) was to "identify barriers to, and opportunities for, more effective communication among scientists and coastal stakeholders about improved monitoring, forecasting, mapping, and other data collection and research regarding long-term changes in U.S. coastlines." The committee determined that this report would have greater value if it not only focused on

[1] There are numerous models of efficient or effective communication. This chapter focuses on effective communication when information is conveyed with a shared meaning or a common language to achieve a goal (Grice, 1975; Granek et al., 2010; NASEM, 2017a).

the Gulf Coast, but also broadly addressed effective communication between scientists and stakeholders with regard to scientific knowledge and understanding, which encompasses the specific information mentioned in the SOT. This chapter identifies the barriers, while opportunities are addressed in Chapter 5. The following sections describe one proven model for effective communication (i.e., boundary organizations) and some barriers from the perspective of stakeholders and scientists.

Over the past few decades a model for more effective communication has emerged, one that involves two-way flows of information among all those involved (Newstrom and Davis, 1986) and is sometimes mediated by a third party such as a boundary organization.[2] In this chapter, boundary organizations are conceptualized broadly as playing an intermediary role between different disciplines and expertise, by facilitating relationships between information producers and users, and integrating user needs with the activities of information producers (Feldman and Ingram, 2009). Boundary organizations create and sustain meaningful links between knowledge producers and users, and are accountable to both. These organizations generally focus on user-driven science, seek to provide a neutral ground for discussion, and help deliver the resulting science to audiences who can use it. Some of the institutions that play the role of a boundary organization in the Gulf Coast include the National Oceanic and Atmospheric Administration (NOAA) Sea Grant network (Sea Grant, 2018), nongovernmental organizations such as The Nature Conservancy (2018), and research institutions and consortia such as The Water Institute of the Gulf (2018) and the Consortium for Resilient Gulf Communities (2018).

Scientists and stakeholders sometimes play a more direct, engaged role in communication, whether they communicate among themselves or with the assistance of a boundary organization (Feldman and Ingram, 2009). Typically, a scientist or a stakeholder who is involved in the practice of communication plays the role of a "boundary spanner" (also called an "intermediary")—an individual working at the edge of different groups who serves to connect those groups with each other. Boundary spanners engage in communication and outreach, whether professionally or more informally, that often serves to process information and represent the voice of outside stakeholders. Boundary spanners also work to engage stakeholders, negotiate relationships, and build connections among the different individuals or groups (e.g., scientists, stakeholders, boundary organizations) (Sandmann et al., 2014). For example, in the Gulf Coast, individuals in organizations such as the U.S. Army Corps of Engineers and initiatives such as the Louisiana *Coastal Master Plan* have played the role of boundary spanner for a number of years.

[2] The Belfer Center for Science and International Affairs (Guston et al., 2000) offers a well-recognized definition of boundary organizations: "Boundary organizations are institutions that straddle the shifting divide between politics and science. They draw their incentives from and produce outputs for principals in both domains, and they internalize the provision and ambiguous character of the distinctions between these domains. It is hypothesized that the presence of boundary organizations facilitates the transfer of usable knowledge between science and policy."

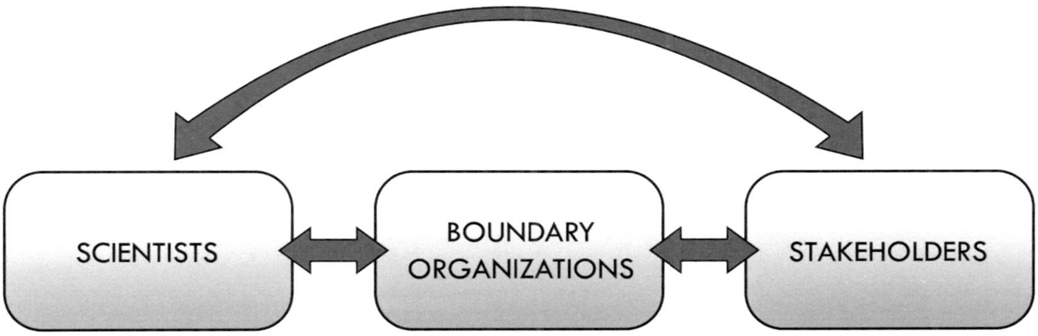

FIGURE 4.1 An idealized two-way communication among all interested parties: scientists, boundary organizations, and stakeholders.

Figure 4.1 shows an idealized representation of effective communication flow among scientists (those who generate data and improve system understanding, whether from the government, academia, or the private sector), boundary organizations (facilitators who serve as brokers, negotiators, and translators between scientists and practitioner stakeholders or the public), and practitioner stakeholders (e.g., city planners, emergency managers, coastal planners). It is important to note that scientists can also be stakeholders and that there may be some overlap among the three groups. Communication may occur across and among all three of these groups (and this is often most ideal), or the pathways may be two-way. For example, when communication occurs directly between scientists and stakeholders, scientists will tend to communicate the latest results that are usable and actionable by stakeholders, while stakeholders communicate their knowledge and understanding, which in turn informs and grounds scientific research.

STAKEHOLDER PERSPECTIVE

As has been mentioned in previous sections, the Gulf Coast is heavily influenced by the energy industry. As such, much of the population along the northern and central Gulf Coast is integrally linked with the energy industry's successes and challenges (see Figure 4.2). This is especially relevant in the face of projected changes to the natural coastal system and underlying feedbacks with the industry infrastructure and operational activities. There is an awareness among the populace that the energy industry needs a highly skilled workforce (Theriot, 2014; Hochschild, 2016), and that the energy workforce is an integral part of a coupled natural-human system. Furthermore, there is an understanding that long-term

FIGURE 4.2 A poster from the annual Shrimp & Petroleum Festival held in Morgan City, Louisiana. This festival exemplifies the integral role that the energy industry plays in coupled natural-human systems in the northern Gulf of Mexico. SOURCE: Shrimp & Petroleum Festival.

physical changes to the Gulf Coast influence both the local population and the energy industry, and that an increase in flood risks affects energy infrastructure, the people who work with that infrastructure, other people in the region, and the infrastructure that they use (LACPRA, 2017). This is a potential opportunity for effective communication.

However, the deep historical linkages and dependencies between the energy industry and surrounding communities can create challenges for effective communication, especially regarding research needed to understand long-term physical changes along the Gulf Coast. For example, several authors have noted that people in Louisiana are simultaneously grateful to the energy industry for providing well-paying jobs in a region that has not always had a strong economy, while also being critical of the energy industry's role in wetland loss and environmental contamination (Theriot, 2014; Hochschild, 2016). Hochschild (2016) describes how these complex employment and social ties impact the flow of information among citizens, stakeholders, and decision makers at public forums in southern Louisiana.

It is within this sometimes contradictory backdrop that the committee discusses more effective communication between scientists and stakeholders. Insights reported here come primarily from in-person discussions at a committee workshop held in September 2017 and from information-gathering sessions during other committee meetings (see Chapter 1 and Appendix B for additional information). Insights presented here are drawn from those discussions and subsequent deliberation within the committee, supplemented by relevant scientific literature. The stakeholders who participated in the workshop and in meetings represent practitioners working on the frontlines of issues related to coastal resiliency and planning at the local and the regional level, including state and local agencies, the energy industry, organizations working with the public, and boundary organizations. These practitioner stakeholders are both producers and consumers of scientific information, as well as translators of that information to other stakeholders, including members of the general public. Stakeholders not present at the workshop but who may have a focused interest in coastal systems include those who fish for a living or for recreation, those who depend on or engage in tourism, and business owners in flood-prone areas, among others. These individuals are likely to be served by practitioner stakeholders and boundary organizations and may engage with scientific information through media. The majority of insights from the literature come from the *Gulf Coast Climate Needs Assessment* (Needham and Carter, 2012), conducted in 2012 by the Southern Climate Impacts Planning Program.

Five main barriers emerged from discussions with stakeholders in the region: resource constraints, limitations on the usefulness of scientific products to support decision making, factors contributing to communication success and failure, difficulties with communication among stakeholders, and challenges for boundary spanners and organizations. Each of these is discussed in further detail in the following section.

Resource Constraints

Four resource constraints—money, time, availability, and expertise—reportedly make the use of scientific information difficult in stakeholder decision making. Lack of financial resources or budgetary constraints can hinder access to information, especially when there is a fee associated with information access (Needham and Carter, 2012). Limited time is another constraint. Given the broad range of activities that practitioners engage in during their work day, it can be difficult to block out the time necessary to sort through available scientific research and interpret data for informed decision making. Even if time were not a constraint, it is difficult for stakeholders to know about all possible types and sources of available information that may be of use to them. In addition, the proprietary nature of some scientific information collected by the private sector (e.g., some data collected by the energy industry or others in the private sector) may prevent access. Lack of expertise may also hamper stakeholder ability to use scientific information to make informed decisions even if they have access. Some stakeholders may not be able to easily discern trustworthy or credible sources, and thus may be ill-equipped to judge the quality of scientific research.

Limitations on the Usefulness of Scientific Products to Support Decision Making

Relevance, timeliness, and usability of information are all central to the extent to which it can be used effectively for decision making. Stakeholders get their scientific information from a variety of federal, state, and local sources. In addition to these sources, stakeholders also draw from local knowledge and citizen science to inform their decisions with respect to resiliency and planning. Stakeholders felt that decision making in the Gulf Coast could be improved if stakeholders had more rigorous, comprehensive information that better characterizes and integrates the ecological, physical, and human components of the coupled natural-human system.

Practitioner stakeholders use scientific products (e.g., decision-support tools, simulations, datasets) to inform their decision making. Products tailored to specific user needs tend to be the most useful, applicable to, and used in stakeholder decision making (NRC, 2009; Dietz, 2013). These could be developed in conjunction with stakeholders or with stakeholder needs in mind from the beginning. Many products, however, are neither directly relevant to stakeholder needs nor vetted for quality and applicability. Many scientific products and tools are not accompanied by sufficient instructions or training on how to apply the provided information to the decision-making process, which means potentially useful information could go unused or be misused. Many stakeholders acknowledge they end up relying on products that are intuitive and easy to use, rather than potentially more sophisticated but complicated tools.

Practitioners also have to be careful that scientific information is not seen as serving the interests of one group over another, which could affect the extent to which stakeholders are willing to engage or use certain data. This was a particular point of focus for Louisiana when developing its *Coastal Master Plan* (LACPRA, 2017). To build trust in the plan's results, the state appointed a number of scientific advisory panels to oversee and review scientific research during plan development, conducted regular public and stakeholder outreach to explain the science behind the plan, made resulting datasets available for public use, and ensured all relevant policy analyses, metadata, and quality control were well documented.

Factors Contributing to Communication Success and Failure

There are many modes of communication among scientists, practitioners, and the general public, ranging from preparing and sharing technical documents and convening outreach meetings to posting on social media and presenting educational displays in public spaces.

Workshop participants expressed the view that communication success can be measured in a number of ways, including whether projects are supported by the larger constituency, even though some conflicts among the parties may exist; whether there is consensus among diverse groups (or at least mutual understanding of others' perspectives); whether conflict is reduced; and whether the public is able to discern among facts, assumptions, and values when making decisions. Some case studies related to communication success with Gulf Coast stakeholders are found in Box 4.1.

A lack of financial resources, the logistical complexity of communication between scientists and stakeholders, difficulty in identifying other relevant stakeholders, and lack of human resources contribute to communication failure. Stakeholders at the Houston committee workshop also reported, specifically, that some members of the public in the Gulf of Mexico do not understand or accept scientific consensus about climate change, which makes it difficult to talk about issues such as sea level rise (as also noted by Needham and Carter, 2012). There is a growing body of literature suggesting the extent to which science, especially politically polarizing topics such as climate change, provides barriers to effective communication. More research is needed to understand how to enhance evidence-based decision making (see, e.g., NASEM, 2017a). Moreover, regional differences between stakeholders and the general public can make it difficult to know how to best translate information for delivery to different groups.

Difficulties with Communication Among Stakeholders

Stakeholders engage regularly and well with local, state, and federal agencies; elected officials; nongovernmental organizations; the private sector (e.g., engineering firms, con-

BOX 4.1
Case Studies of Gulf Coast Stakeholder Communication

Previous regional studies that aimed to enhance stakeholder communication focused on understanding how personal experiences influence perceived (not actual) impacts, paying close attention to communication content and delivery modes, and including those with extreme views, with a goal toward mitigating them.

For example, in a transdisciplinary research project called the Ecological Effects of Sea Level Rise in the Northern Gulf of Mexico (DeLorme et al., 2016), there was a focus on what made communication between scientists and stakeholders successful (or not). Important factors included close attention to communication content and format, delivery, and outcomes throughout the lifetime of the project, and being open to improvements and refinement as necessary.

When studying the results of a citizen science project on rapid marine invasion by lionfish in the northern Gulf of Mexico, Scyphers et al. (2015) found personal observations of invasive species strongly influenced perceived impacts and positively influenced support for initiatives.

A multicriteria decision analysis focused on coastal resilience in Mobile Bay, Alabama (Bostick et al., 2017), brought together approximately 30 stakeholders from local and regional emergency and environmental management, the port industry, and public works with the U.S. Army Corps of Engineers subject-matter experts, with a goal of keeping those with extreme views in the discussion so that they would be able to gain an appreciation of other views and to see if scenarios of concern would influence the priority order of decision making when the group came to consensus. "Increased storm frequency" and "greater urbanization and development" had the largest influence on changing priorities, while "increased storm severity" and "decline in shipping industry" had the least.

sultants); and the public. Generally, interactions with academics (both within the Gulf Coast region and beyond it) were considered to be better when there was a two-way flow of information between scientists and stakeholders rather than a one-way flow from scientist to stakeholder or vice versa (e.g., Beierle, 2002; Kirchhoff et al., 2013). However, there is a need for greater interaction with industry and with vulnerable populations.

Based on workshop and information-gathering meeting discussions, there appears to be limited information sharing between stakeholders and some parts of the private sector (most notably, the energy industry, although other sectors mentioned included the insurance and reinsurance industries, finance sector, and the military) on topics such as industry assets, risk mitigation and communication, and emergency planning. Communication among stakeholders and industry is complicated for several reasons. First, the energy industry is not homogenous and represents many large, complex organizations, so it is not always clear to whom stakeholders should be directing their inquiries or how to develop a strategy for communication. Second, industry representatives may have reasons for not sharing data, including the desire to prevent their competitors from gaining an economic advantage, financial disclosure laws or regulations, impacts of disclosures on financial markets, and the possibility of litigation.

Stakeholders often have difficulty reaching out to vulnerable populations, such as rural and indigenous communities, low-income communities, people who speak different languages, communities with low literacy levels, and disabled or elderly people. Box 4.2 discusses an approach taken by the State of Louisiana to address this issue.

Challenges for Boundary Spanners and Organizations

Boundary spanners and boundary organizations face three main challenges. The first challenge is obtaining approval or acceptance from local- or state-level officials and community members to participate in community engagement, which, if achieved, can help community leaders see the value of engaging with boundary spanners or boundary organizations. The second challenge is establishing true two-way flows of information between scientists and stakeholders, especially if participants on either side hold more parochial perspectives about their roles and the provision of scientific information. The last challenge

BOX 4.2
LA SAFE: An Example of Involving Vulnerable Stakeholders in Adaptation Planning

Louisiana's Strategic Adaptations for Future Environments (LA SAFE) is an outreach initiative led by the state government that includes a wide range of practitioner and other stakeholder partners.[a] LA SAFE takes a "people-driven approach" to engage with communities in six parishes at risk of coastal flooding and land loss, with a goal to develop an adaptation plan that is sensitive to the culture and needs of the people who will be affected. LA SAFE takes a holistic approach by explicitly engaging with stakeholders on topics related to demographics (e.g., social vulnerability, population shifts, age and diversity, housing and income), economics (e.g., tax base, real estate, jobs and industry, fishing and farming, commuter patterns), social issues (e.g., health and safety, organizations, education, culture, politics), and environment (e.g., pollution, subsidence, land change, sea level rise, habitats, and risk). Community stakeholders are recruited mainly through nonprofit organizations that work on the ground and that already have deep roots in hard-to-reach, immigrant, and vulnerable and rural communities. Recruitment efforts involve some combination of phone calls, flyers, and door knocking, among other methods. LA SAFE develops and maintains a relationship with these community stakeholders and provides them with feedback on what other community members are saying, including summarizing results and presenting them to stakeholders in meaningful and understandable terms, and also provides food and childcare to make it easier for community stakeholders to participate in meetings.

[a] More information about LA SAFE is available at: https://lasafe.la.gov.

is that the process of communication necessitates the involvement and coordination of multiple entities and individuals, which can make progress slow or difficult.

SCIENTIST PERSPECTIVE

Despite growing recognition over the past few decades of a need for improved communication between scientists and stakeholders, such communication often remains limited to scientists (whether in academia, industry, or as a practitioner) who are either inclined to do so, have some level of job security (e.g., tenured faculty), or both (Jacobson et al., 2004). Four main barriers are discussed in the following section: institutional and interpersonal barriers, trust, "outsider" status, and underuse of science communication tools.

Institutional and Interpersonal Barriers

Weerts and Sandmann (2008) discuss factors that are barriers or facilitators to effective communication at institutions such as research universities. Institutional barriers arise for a variety of reasons, including leadership (e.g., chairs, deans, provosts, supervisors, senior colleagues) who may discourage a perspective that views community members as stakeholders or learning partners, and thus oppose community engagement. A personal barrier exists if a scientist does not see value in communication and mutual learning with stakeholders

Some scientists hold the view that stakeholders, especially members of the general public, do not act rationally (Ariely, 2009). Therefore, engaging with stakeholders is pointless because the presentation of scientific facts will be met with fear, dismissiveness, misunderstanding, or worse. Other research suggests that people do not in fact act irrationally (Fischhoff, 2013; Fischhoff and Scheufele, 2013); rather, they react to information in view of their unique circumstances, understandings, ideologies, and so on. Other scientists view communication activities, such as engaging with stakeholders, as a form of advocacy, which they do not see as their role in society (Lackey, 2007). Engaging the community may force scientists to either take a position that undermines their credibility and legitimacy or influences their ability to be objective, which could compromise their research.

Trust

Trust and confidence may also limit effective communication between scientists and stakeholders. Research in risk communication suggests that trust (comprising perceived competence, objectivity, fairness, consistency, and faith) and confidence (an enduring experience of trustworthiness over time) are necessary for effective communication (Renn and Levine, 1991). Both trust and confidence are necessary components when assigning credibility to a source of information or an organization. Trust in media, industry, experts, government, and nongovernmental organizations is on the decline (Harrington, 2017). This lack of trust and confidence can make it very difficult for communication among scientists, boundary organizations, and stakeholders to be effective.

"Outsider" Status

Another factor that limits communication is that scientists from outside the Gulf Coast may be considered as "outsiders" by "insider" stakeholders. Research in social psychology and sociology suggests that relationships between ingroup and outgroup members tend to be marked by discrimination, whether that be favoritism toward insiders and an absence of favoritism toward outsiders or being dismissive of outsiders if they are seen as a threat or hindrance (Turner et al., 1987). This dynamic can make it very difficult, if not impossible, to engender trust among groups, let alone build effective communication. Indeed, this has been observed across a wide range of domains involving communication among stakeholders from management of natural resources (De Nooy, 2013) to decision making about energy development (Wong-Parodi et al., 2011).

Underuse of Science Communication Tools

The emerging field of science communication provides a broad set of tools for more effective communication between scientists and stakeholders (e.g., NASEM, 2017a), and more established fields of participatory action research, citizen science, and coproduction offer suggestions for how to establish better communication among interested groups (e.g., Whyte, 1991; Hassol, 2008; Bonney et al., 2009; Silvertown, 2009; Dilling and Lemos, 2011; Somerville and Hassol, 2011). The purpose of these tools is to facilitate lines of communication through which scientists can learn from stakeholders and stakeholders can learn from scientists. For example, participatory research was recently used to study the condition of infrastructure in Houston (Hendricks et al., 2018). Despite this richness, many scientists may not have the resources, tools, expertise, incentives to use, or awareness of the tools available to enhance effective communication, nor those frameworks for effectively developing those relationships that foster communication.

There is often an information gap between scientists and stakeholders, when each may hold expertise in their own domain areas and there is a need to share information. Some scientists believe that one way to bridge this information gap is through the provision of more information (Ben-Haim, 2001). Research suggests, however, that more information is not necessarily helpful; rather, providing the information that people need at the appropriate level is more effective (Bruine de Bruin and Bostrom, 2013; Fischhoff and Scheufeleb, 2013; Wong-Parodi et al., 2013).

Decision making under deep uncertainty (DMDU) methods are often used to support a participatory "deliberation with analysis" approach (NRC, 2009). This approach to decision analysis uses scenario analysis and interactive visualizations to facilitate conversations among decision makers, stakeholders, and residents. Critically, different viewpoints (e.g., assumptions about future conditions, preferences across different goals) can all be captured in the same analysis, allowing for an inclusive framework that seeks to highlight key tradeoffs, policy-relevant scenarios, and tipping points for deliberation rather than mask them through probabilistic assumptions or "black box" modeling choices.

MEDIA AND COMMUNICATION

Media formats play an important role in communicating issues regarding long-term coastal change to the public, decision makers, and stakeholders across the Gulf Coast. Some traditional news outlets (e.g., television, radio, print, and their online components) have dedicated coastal reporters, and as such, coastal issues are often in the news. There have been long format pieces (e.g., documentaries) about the region and the interactions between the landscape and the energy industry. The widespread presence of online weather and hydrographic data and models are readily being used by people to evaluate their own flood risks, as well as to enhance fishing trips and to ensure safety during recreational boating activities.

Both traditional and emerging formats can present barriers to communication. For example, some journalists may have a hard time evaluating key background information—which can be particularly problematic in the Gulf Coast given that some regions are educationally underserved. Additionally, the complexity of relationships (ecological, physical, and social) that exist across regions between the landscape and coastal development are often hard to summarize in many media formats that often require relatively short pieces. While digital and social media is becoming more common across the region, high poverty in some places can limit access to digital tools. For example, in 2015, 74 percent of U.S. households with an annual income less than $30,000 reported having access to the Internet, as opposed to 97 percent of households with an annual income greater than $75,000 (Perrin and Duggan, 2015).

Both traditional and emerging digital media formats present opportunities to communicate with the public, stakeholders, and decision makers. For traditional media, the presence of dedicated coastal reporters offers prospects for more communication. The cultural economy that exists across the Gulf Coast also offers opportunities to enhance communication concerning people and the landscape. Furthermore, social and other emerging digital media platforms offer chances for better communication and have proven useful particularly during environmental events and disasters.

CHAPTER 5

Defining a Research Agenda: Priorities to Increase Understanding of Long-Term Coastal Zone Dynamics

The establishment of the independent, science-based National Academies of Sciences, Engineering, and Medicine's Gulf Research Program (GRP) as part of legal settlements with the companies involved in the 2010 *Deepwater Horizon* oil spill presents a unique and timely opportunity for the creation of a sustained, integrated research program to increase understanding of long-term coastal dynamics. The development of such a holistic effort, with coordination and integration across multiple disciplinary research foci, has the potential to improve human well-being in the coastal zone along the Gulf Coast and around the world. The research agenda defined in the following section leverages the scope and autonomy of the GRP with suggestions for a long-term, integrated body of research that federal and state funding agencies would find hard to achieve due to their mission-oriented natures.

In developing this research agenda's vision, the committee was guided by three criteria: the independent and long-term nature of the GRP, the research gaps identified in Chapter 3, and the significance of these research gaps at temporal and spatial scales relevant to the Gulf Coast.

VISION AND CRITICAL AREAS OF RESEARCH

In response to the first part of the Statement of Task (SOT) (see Box 1.1), the committee identified key high-priority research gaps that reflect the need for better understanding of critical physical, ecological, and human processes in the coastal system (see Chapter 3). The second part of the SOT asked the committee to define the essential components of a research and development program to address these research gaps. To do this, the committee developed a vision that addresses the critical areas of research. This vision is "*to understand and predict the feedbacks and interactions among the physical, ecological, and human components and the resulting evolution of the coupled system along the U.S. Gulf Coast, in the context of both human and climate drivers.*" These critical areas encompass the research gaps identified in

Chapter 3 and provide some overarching context for a potential research and development program.

The goal of identifying the critical research areas is to provide the ability to project the future state of the Gulf Coast, given changing environmental stressors and a range of policy options related to mitigation or adaptation strategies. Such a capability will enable improved probabilistic assessments of future conditions (where plausible) and can also be used to identify key scenarios, including those situations when the uncertainty cannot readily be quantified over these timescales. To achieve this goal, a more complete understanding of the interactions and feedbacks between the natural (physical and ecological) and human systems is needed. Three critical areas for research are identified below:

- **Critical Area 1:** How will coastal landforms and coastal ecosystems along the Gulf Coast respond to rapidly changing conditions (both natural and human induced), especially given the expectation for continued relative sea level rise acceleration?
- **Critical Area 2:** How will human settlement and economic activity along the Gulf Coast respond to evolving coastal landforms and ecosystems under rapidly changing conditions?
- **Critical Area 3:** How can improved understanding of both near- and long-term evolution of the Gulf Coast coupled natural-human system be applied to inform stakeholder decisions made at local, state, and regional scales? How does the coupled natural-human system evolve when decision making is updated as scientific understanding advances?

The dominant processes, interactions, and feedbacks for all three critical areas vary by the timescales of the physical and ecological drivers of change for the natural system, as well as the motivators for human response and decision making. While these timescales range from nearly instantaneous to several thousands of years, the relevant timescale for this report (as noted in Box 1.2) is 10 to 200 years, subdivided into near-decadal (10 to 50 year) and decadal-century (50 to 200 year) scales.

As presented in Chapter 2, evolution of the Gulf Coast over a near-decadal timescale (approximately 2030 to 2070) is likely to be dominated by feedbacks and interactions associated with episodic events. These events include not only inundation and wave-driven flooding caused by tropical cyclones, but also smaller magnitude events such as cold fronts, winter storms, and seasonal tides, all exacerbated by sea level rise. Within the same timeframe, Gulf Coast evolution will also be influenced by population growth or decline (including individual and household relocation decisions), regional and national economic trends, local coastal development, natural resource extraction, and ecosystem conservation and restoration initiatives.

Coastal evolution at a decadal-century (50 to 200 year) timescale is likely to be dominated by the influence of relative sea level rise. At this timescale, accelerating sea level rise will lead directly to significant changes along the Gulf Coast, including large-scale alterations of coastal landscapes and ecosystems and potential abandonment of at-risk communities. Global climate change and sea level rise impacts could lead to secondary impacts on the coupled natural-human Gulf Coast system.

GUIDELINES FOR A RESEARCH AGENDA

Informed decision making and planning through both the near-decadal (10 to 50 year) and the decadal-century (50 to 200 year) timescale will require improved understanding and capability to project relative sea level rise and changes in episodic events, the coastal responses to these processes, and the ways that humans respond to and influence these processes. **This type of research and development program will be most successful if it focuses on the interactions and feedbacks critical to evolution of the coupled coastal system; is carried out by collaborative, multidisciplinary research teams; supports comprehensive, Gulf Coast-wide, fully integrated modeling and observational efforts; encourages longitudinal, multi-decadal research; produces easily accessible observational data and model results; and is coordinated at a high level.**

Instituting a mechanism for rapid and easy sharing of data and models across program participants and beyond, as well as the use of data assimilation to improve model skill, would also enhance the success and reach of research efforts. Essential components of this long-term research agenda are discussed in turn in the following section. Examples of some programs that include these components are discussed in Box 5.1.

Focus on Interactions and Feedbacks Critical to Evolution of the Coupled Coastal System

A research and development program developed to understand and project future coastal evolution would benefit from a focus on the interactions and feedbacks between and among the components of the coupled natural-human coastal system. While it is very useful to study specific aspects of the natural or human systems (discussed in greater detail in Chapter 3), integrating this research is likely to provide the greatest leaps in understanding.

Collaborative and Multidisciplinary Research Teams

Many of the current gaps in understanding discussed previously in this report (see Chapter 3) are complex and are unlikely to be addressed by a single discipline; rather, collaborative research teams involving multiple disciplines across the natural and social sciences will be needed. These teams are critical for a more integrative understanding of the

BOX 5.1
Examples of Successful Large-Scale Research Programs

The Chesapeake Bay Program is a coordinated, multi-state effort to collectively manage and improve the ecological health of this estuary. As part of this effort, the program supports a long-term research agenda to address interrelated coastal and human processes across the region. Originally established in 1983 by an agreement among the U.S. Environmental Protection Agency, states, and local decision makers, the program enables regional coordination to help reduce chronic water pollution challenges and support ecosystem restoration in and around the Chesapeake Bay (Chesapeake Bay Program, 1983). The list of partners and commitments has grown in recent decades, culminating in a 2014 Chesapeake Bay Watershed Agreement that includes 10 interrelated goals, such as improved water quality, healthy watersheds, sustainable fisheries, and climate resiliency (Chesapeake Bay Program, 2014).

Of particular note are the program's long-term commitment to regional, coordinated monitoring and simulation modeling. For example, partners monitor and share region-wide datasets describing freshwater flows and contaminant loadings into the Chesapeake Bay, water chemistry (e.g., temperature, salinity, dissolved oxygen), and benthic species and fin/shellfish abundance (Chesapeake Bay Program, 2018b). In addition, program scientists and partners have built and updated a suite of computer models that are together used to project current and future land use, hydrology, and water quality in the tributaries and in the Chesapeake Bay itself. This suite of models, now in its sixth iteration (i.e., "Phase 6"; Chesapeake Bay Program, 2018a), is also used to project the pollution-reduction benefits from various best management practices implemented across the watershed. Monitored data are used to calibrate and validate model results across the region, and this sustained and interrelated monitoring and modeling approach can serve as a blueprint for the program of research described in this chapter.

The Gulf of Mexico Research Initiative[a] (GOMRI) was established in 2010 after the *Deepwater Horizon* oil spill. Its purpose is to conduct a 10-year research program on oil spills and their aftermath on the Gulf of Mexico ecosystem, as well as the impact on shorelines and public health. The governing body is the GOMRI Research Board, which has representatives from the five Gulf states. The Research Board identified five research themes and issued a Request for Proposals for the creation of research consortia (of four or more institutions) to address one or more of the themes. Fifteen consortia have been funded to date.

GOMRI researchers must make data obtained during the project publicly available. The mechanism to achieve this goal was the establishment of the Gulf of Mexico Research Initiative Information and Data Cooperative, which created infrastructure for researchers to archive diverse datasets and provided a portal for the public to access these data.

---

[a] More information on GOMRI is available at: http://www.gulfresearchinitiative.org.

societally relevant, user-inspired questions that can drive policy and decision making, but may also require more flexible or innovative approaches to fund and support their research.

Comprehensive, Gulf Coast-Wide, Integrated Observational and Modeling Efforts

Understanding and projecting the long-term evolution of the Gulf Coast will require both observational and modeling endeavors. Coordinating and integrating observational and modeling efforts will significantly amplify the gains that would otherwise be achieved by observation or modeling alone. Observations provide a means for assessing and quantifying how the coastal system is changing through time, while also providing inputs to models and data critical to the development of realistic model parameterizations. Modeling endeavors provide guidance to observational programs regarding critical variables and field relationships to be measured, as well as the ideal frequency for data collection. Model simulations can also provide projections of coastal change that can be compared with longitudinal observations. The integration of observational and modeling programs, ideally through an iterative design, facilitates the development of targeted and adaptive observational programs, as well as the continued development of models and improvements in modeling skill.

Essential components of a coordinated, integrated, and iterative observational and modeling program would focus on furthering the understanding of the long-term co-evolution of the hydrodynamic, morphological, ecological, and human systems, including consideration of wave circulation, sediment transport, coastline evolution, land elevation, vegetative processes, water level, water movement, water quality, demographic changes, measures of economic activity, and human decision-making processes, responses (e.g., abandon, migrate, harden, nourish), and impacts (e.g., on fisheries).

A viable approach to developing a Gulf Coast-wide observation and monitoring program would be to enhance and extend existing monitoring programs, while also implementing new and emerging in situ and remote sensor networks. For instance, new and existing observational programs could be employed with in situ physical and chemical sensors, remote sensing, ecological collections, and data surveys. Data are also needed to assess the impacts of sea level rise and changes in air temperature, rainfall, and rainfall frequency.

Similarly, a Gulf-wide coastal evolution model could build on existing models of specific coastal processes, but also allow for new integrated approaches that examine the co-evolution of the natural-human system. Such a modeling framework would enable projections of coastal evolution as well as broader economic, cultural, and societal changes under different future climate and management scenarios. Ensemble simulations may provide an important means for evaluating the uncertainty of future projections. Data-assimilation techniques can aid in improving model skill and provide guidance for the collection of observations. Implementing ensemble modeling and data-assimilation approaches, where appropriate, will likely enhance the success of a research and development program designed to address the priority research gaps identified in this report.

Longitudinal and Multi-Decadal in Scope

There are few research opportunities to track change to the coupled natural-human system over long temporal scales, and even fewer opportunities in which targeted modeling efforts can be coordinated to make projections for future decades and validated with repeated observations over time. A research and development program that intentionally takes the long view, iteratively making projections and observing change over time, has the power to transform understanding of coastal evolution and to revolutionize the ability to project coastal change in the face of uncertain future conditions. Longitudinal[1] observational, experimental, and monitoring programs can facilitate synthesis efforts aimed at tracking the drivers of change, quantifying patterns, and identifying cascading impacts through the system. Though longitudinal studies are emphasized here, shorter-term (e.g., at the event scale or annual scale) research on physical, ecological, and human processes will also contribute to longer-term understanding. As the Gulf Coast experiences changes in sea level and climate, new questions will emerge regarding the patterns and drivers of change and their impacts on the coupled system. For this reason, a program that is longitudinally focused with integrated modeling and observational components, but that is also adaptive and responsive, will be best poised to provide maximum benefit.

Easily Accessible, Regularly Updated Observational Data and Model Results

To the extent possible, making data publicly available in real time and archiving them in accessible databases will extend the utility of data beyond scientific research and assist emergency managers, planners, other researchers, and decision makers with adapting and responding to changing environments or emerging disasters. Likewise, developing models using open source platforms and sharing newly developed components soon after they are vetted will extend their impact beyond the originators, facilitating their use and application.

High-Level Coordination

Management of a research and development program that includes efforts that are Gulf Coast-wide and longitudinal, contains highly integrated modeling and observational components, and emphasizes interactions and feedbacks between the natural and human

[1] Longitudinal studies collect repeated observational data on cross-sectional units (e.g., individuals, groups of people, cities) over time.

coastal systems will require intentional, consistent, and careful administration. Such high-level coordination can be achieved in a variety of ways, including close collaboration across co-developed projects, sharing of project personnel among related projects, and supporting projects that include significant data synthesis and integration aspects. To fully realize the integration of modeling and observational efforts, projects will need to be designed from the start with the intention to integrate, iterate, and synthesize information.

RESEARCH PRIORITIES BY CRITICAL AREA

Critical Area 1: How will coastal landforms and coastal ecosystems along the Gulf Coast respond to rapidly changing conditions (both natural and human induced), especially given the expectation for continued relative sea level rise acceleration?

Among the wide range of stressors that affect the Gulf Coast (e.g., changing climate, land use, sediment management), low-elevation coastal landforms and ecosystems are particularly sensitive to the acceleration of the rate of global sea level rise. As discussed in earlier chapters, relative sea level rise rates along the western half of the Gulf Coast will be higher than average global rates due to subsidence. The need to understand both the oceanic component of sea level rise (Research Gap 1) and subsidence (Research Gap 2) with some geographic and temporal specificity is critical to anticipate and plan for effects of rapid relative sea level rise on Gulf Coast landforms and ecosystems.

High rates of relative sea level rise, combined with both historic and potential future reductions in sediment supply (Research Gap 4), threaten coastal landforms and the ecosystems they support. Of critical importance along the Gulf Coast are the evolution and fate of low-lying features, especially barrier islands, river deltas, and wetlands (e.g., swamps, marshes, mangroves), which are often tightly interconnected. Despite a general recognition that large-scale morphological changes are likely, the understanding of—and the ability to predict—where, when, and how changes will occur is limited (Research Gap 6), particularly given the potential impacts of episodic extreme events in combination with relative sea level rise (Research Gap 3) and the effects of human modifications to the coastal zone (Research Gaps 9 to 12). It is critically important to project when, and under what conditions, coastal evolution transitions from being dominated by episodic events (at the near-decadal [10 to 50 year] scale) to being dominated by relative sea level rise (decadal-century [50 to 200 year] scale). For example, under which (natural and human-modified) conditions will barrier islands no longer be able to migrate landward rapidly enough to avoid disintegration and drowning (Research Gap 6) and under which conditions will coastal wetlands no longer be able to sufficiently accrete vertically or expand laterally to avoid becoming submerged (Research Gap 7)? How will sediment supply (Research Gap 4), including the effects of human interventions (Research Gaps 9 to 11) and ecosystem conservation initiatives

(Research Gap 8), affect the evolution of deltas as sea level rises (Research Gap 6)? How will changes to, or the loss of, coastal landforms and ecosystems affect surge-induced flooding, tides, and erosion farther inland (Research Gap 3)?

These types of dramatic changes will transform the natural environment in ways that will influence the built environment (Research Gap 10), patterns of human migration (Research Gap 11), feedbacks between the human and natural systems (Research Gaps 6 and 9 to 12), economic activity, and how, when, and to what extent humans continue to inhabit the Gulf Coast in the future. Understanding and projecting the evolution and fate of Gulf Coast landforms and ecosystems will therefore not only require a better understanding of key processes, such as feedbacks between sediment transport and vegetation (Research Gaps 5 and 6), improvements in coastal zone sediment transport formulations (Research Gap 5), and modeling capability (Research Gap 6), but it will also require identification, exploration, and simulation of the interactions and feedbacks between the natural and human systems that will be most important in determining the future fate of coastal landforms and ecosystems, and therefore, human habitation in this region.

Critical Area 2: How will human settlement and economic activity along the Gulf Coast respond to evolving coastal landforms and ecosystems under rapidly changing conditions?

Anticipated changes to coastal landforms and ecosystems will significantly impact human settlement and the built environment. A deeper understanding of the interactions and feedbacks between the built environment and the natural system (Research Gaps 9 and 10) and between human migration and the natural system (Research Gap 11) can inform models of the coupled natural-human system (Research Gap 12) to improve long-term decision making.

There is considerable uncertainty about the extent to which Gulf Coast residents and stakeholders will attempt to maintain coastal landforms and defend existing infrastructure (Research Gap 9) versus allowing land and habitat loss to occur. Some barrier island communities (e.g., Dauphin Island, Alabama) have been defended repeatedly, and some barrier island settlements may ultimately cease to exist. Some communities could find new ways of living along the Gulf Coast, including potential approaches that currently seem radical, such as mobile rather than fixed residences.

As people migrate (Research Gap 10) or choose to remain in place in response to changes in coastal landforms (Critical Area 1), urban and rural settlement patterns along the Gulf Coast could change. More people may live in cities that have coastal engineering infrastructure, while rural communities without hazard-mitigation approaches may struggle to remain in place, except for communities of means that can afford defensive approaches. Moreover, settlement patterns may change differently in areas of the western and central

Gulf Coast, where relative rates of sea level rise are greatest and social vulnerability is highest, compared with areas along the Florida Gulf Coast, where relative sea level rise rates and social vulnerability are more modest. Major Gulf Coast cities, such as Houston, New Orleans, and Tampa, could see populations becoming more concentrated in areas with relatively higher elevation, although there is uncertainty about whether populations in these cities will grow or contract as a result of migration and other changes along the Gulf Coast.

All of these changes, interactions, and feedbacks will affect the availability of the resource base (e.g., labor, infrastructure, natural resources) that drives key economic sectors. For example, land loss will influence whether energy infrastructure is abandoned, whether it is modified to adapt in place and continues to operate, or whether it is relocated farther inland or to higher elevations. Tipping points, such as extensive marsh losses or large-scale barrier island disappearance, will affect nursery grounds for fisheries, associated fish stocks, commercial and recreational landings, and the fishing and fish processing economy. Coastal land loss will also impact recreational opportunities and the tourism industry, whether directly through losses in rental properties or indirectly through effects on fish stocks or infrastructure that supports tourism. Expected changes in human settlement and economic activity will depend on the timescales of coastal evolution. At the near-decadal timescale of 10 to 50 years, human demographics could change in response to episodic storms and nuisance flooding. The built environment could change with increased adoption of coastal engineering approaches or with modifications to existing housing stock that would allow people to adapt in place. At decadal-century scales beyond 50 years, there is potential for even bigger change. One possibility is large-scale human migration in response to unprecedented relative sea level rise and land loss. Another example is how the commercial and recreational fishing industries would adapt to major losses in nursery habitat if continued wetland loss and relative sea level rise led to the loss of barrier island systems. Moreover, the built environment could change dramatically as new adaptation strategies and technologies emerge and ways of living in the coastal zone evolve.

Critical Area 3: How can improved understanding of both near- and long-term evolution of the Gulf Coast coupled natural-human system be applied to inform stakeholder decisions made at local, state, and regional scales? How does the coupled natural-human system evolve when decision making is updated as scientific understanding advances?

Research carried out under Critical Areas 1 and 2 will enable improved understanding of the most significant interactions, feedbacks, and drivers of change that influence the long-term evolution of the coupled natural-human coastal system (Research Gaps 1 to 5, 7, and 8). Through integrated observational and modeling efforts, it will also improve the overall understanding of how the Gulf Coast coupled system is likely to evolve in the future (Research Gap 6). Research and applications will be required to connect scientific insights with decision making at the local, state, and regional levels throughout the Gulf Coast. This translation of understanding can help enable iterative decision making that is informed by better understanding of potential short-term and long-term consequences (Research Gaps 9 to 12) and may enable courses of action that were not previously available or considered.

This could entail, for instance, evaluating the potential benefits, costs, and tradeoffs of proposed courses of actions using newly developed coupled modeling systems, applying decision making under deep uncertainty methods (Chapters 2 and 4) to help decision makers identify more robust and adaptive coastal management plans, or using integrated observational data to identify critical tipping points that would suggest switching from one course of action to another. Furthermore, outcomes from research and applications will provide additional insights to guide research under Critical Areas 1 and 2. For example, during the decision-making process, new questions may emerge, and these, in turn, could help inform and prioritize additional basic research pertaining to the coupled natural-human system.

In addition, new scientific understanding can build on the opportunities for communication and outreach identified elsewhere in this report. Sufficient science-based understanding of the risks, benefits, and uncertainties of available options is needed to make reasoned decisions in accordance with values and preferences. Meeting the challenges faced by the Gulf Coast in the future requires that stakeholders have access to useful and usable scientific information (Opportunities 1 and 2 in the following section), and that stakeholders and the public are informed about where such information resides and how to access it. Meeting future challenges also requires creating an environment in which stakeholders and scientists have the resources and incentives to share information (Opportunities 3 to 5) and promote learning. Finally, it entails recognizing limitations and constraints to translating and sharing information and then creating opportunities to overcome these barriers (Opportunities 3, 6, and 7).

BARRIERS TO AND OPPORTUNITIES FOR EFFECTIVE COMMUNICATION

Addressing the research gaps identified earlier in this chapter will substantially advance understanding of the coupled natural-human system of the Gulf Coast and help identify salient feedbacks between humans and their environment in the face of climate change. This, in turn, will constitute crucial information for the development of policies toward a resilient and sustainable Gulf Coast. Implementation of research products into actionable policies necessitates effective communication and collaboration between stakeholders and scientists. In Chapter 4, the report identified specific barriers to effective communication. In this section, opportunities to overcome those barriers are presented.

Stakeholder Perspective

Barrier 1. Financial constraints, information availability, time, and expertise represent a barrier to effective communication. These factors make it difficult for stakeholders to know about, obtain, find, work with, and interpret information and data in a way that allows them to incorporate science into decision making.

Opportunity 1. Targeted funding opportunities that would allow practitioners to obtain data and to hire staff with the expertise and dedicated time to interpret scientific information would facilitate the use and application of available scientific information by other stakeholders. Alternatively, or in addition, the development of a Gulf Coast-wide repository of scientific information managed by well-informed staff who can provide support for stakeholders would be a valuable resource that would help facilitate the incorporation of science into decision making.

Barrier 2. Many scientific products that are intended to help inform decision making are not tailored to stakeholders' specific needs. As a result, the applicability of these products (e.g., tools, data, information) is not clear to stakeholders, who are then less likely to use them for decision making. Furthermore, many scientific products are not accompanied by sufficient instructions or training on how, why, or when to apply the provided information to the decision-making process; the information thus may go unused or may be applied inappropriately. Additionally, some products may be seen by stakeholders as serving the interests of one group over another and thus may not be seen as appropriate for decision making.

Opportunity 2. When developing products that are intended to inform decision making, scientists should be encouraged to engage substantively with stakeholders from the development to the delivery stage. Such an approach can create scientific products that are more likely to be effective and immediately applicable and may help to allay concerns over whether data are serving some needs over others. To encourage stakeholder involvement, solicitations for research programs might include a requirement for substantive and early engagement. Boundary organizations assist in facilitating this type of engagement, and including incentives for their involvement would further improve communication. The degree to which scientific information is used effectively could be further improved by streamlining and guiding the process by which stakeholders identify and access the information they need. Development of an innovative catalog of products would improve the abilities of stakeholders to access and apply these tools in their decision-making activities. Combining this effort with staff support (see Opportunity 1) would facilitate this process.

Barrier 3. The size and complexity of the energy industry, as well as apparent limitations to information sharing, present a barrier to effective communication between the energy industry and other stakeholders.

Opportunity 3. Create an incentive structure that fosters information sharing between the energy industry and other stakeholders, as well as protocols for how to engage more effectively to facilitate information sharing. This process could be facilitated by a third party such as a boundary organization.

Barrier 4. Limited financial and human resources, logistical complexity, and difficulty in identifying all relevant stakeholders, as well as skepticism, lack of understanding, or lack of trust by one or both parties, can make it difficult for practitioner stakeholders to communicate effectively with members of the general public, including vulnerable populations.

Opportunity 4. Boundary organizations can play a key role in facilitating trusting relationships among community members, practitioners, and scientists, allowing for more effective engagement. Advisory committees comprising members of relevant stakeholder groups, including vulnerable or underserved groups, could serve as representatives for their communities and could help identify strategies for more effective communication and engagement.

Barrier 5. There can be difficulties in establishing two-way information flow between scientists and stakeholders, due to one or both parties failing to see the value of communication. Moreover, there are challenges involved in coordinating diverse entities and individuals for any particular research effort, especially when there are numerous people and groups involved.

Opportunity 5. Role-playing exercises may help ensure that scientists, stakeholders, and others see the value of two-way communication. Efforts can be made to demonstrate the effectiveness and value of community engagement through case studies and storytelling, as a first step toward further engagement. Clear lines of communication, chain of command, and protocols, as well as the involvement of boundary spanners or boundary organizations, may facilitate the coordination of stakeholders and scientists in concerted efforts and will help participants feel involved and useful and have a sense of ownership.

Scientist Perspective

Barrier 6. Scientists' engagement with stakeholders can be limited by competing demands on time and by the relative importance placed on this engagement, in terms of promotion and professional recognition. In addition, scientists are often not trained in speaking to public audiences or engaging with stakeholders. They may also not be equipped to efficiently transfer knowledge or to provide appropriately tailored information to stakeholders.

> **Opportunity 6.** Strong relationships, collaborations, and clear communication between scientists and stakeholders help produce scientific results that are most applicable to coastal decision making. To help facilitate the development of key relationships, funding programs could provide funds for engagement and knowledge-transfer activities and consider ways to incentivize collaborations between scientists and stakeholders via boundary organizations and other boundary spanners. Notably, there is an opportunity for extension faculty associated with sea- and land-grant programs to play a prominent role in future engagement and knowledge transfer between scientists and stakeholders. There may also be opportunities to offer training for scientists on effective communication and collaboration with stakeholders. Involving university leadership in these opportunities may also increase interest in and support for scientist engagement with stakeholders.

Barrier 7. Scientists working on or wanting to engage in research relevant to the Gulf Coast but who are not from or based there may feel limited by their "outsider" status when attempting to engage with stakeholders. They may have concerns about whether their information and expertise will be dismissed, especially if the information is viewed as contrary to deeply held stakeholder views.

> **Opportunity 7.** Funding programs that focus on Gulf Coast-related research could encourage and facilitate collaborations among regional scientists (especially those with well-established relationships with stakeholders) and those from outside the region with complementary interests and expertise. To progressively build trust and forge strong collaborations, workshops, personnel exchanges, and symposia could be used to initiate communication and discussions among Gulf Coast stakeholders, Gulf Coast-based scientists, and scientists from outside the Gulf Coast who have relevant research interests and expertise.

The physical and ecological systems, people, and economy in the Gulf Coast are inextricably linked. The research agenda presented here could lead to advances in understanding of the natural and human factors that combine, interact, and feed back to influence coastal evolution, as well as coastal communities and infrastructure. It also underlines the importance of effective communication between scientists and stakeholders in promoting informed decision making in the coastal zone. While changing environmental conditions present challenges to coastal communities, there is also a great opportunity for groundbreaking research and innovation, which may lead to a re-envisioning of what it means to live along the Gulf Coast.

References

Aldrich, D.P. 2012. *Building resilience: Social capital in post-disaster recovery.* Chicago: University of Chicago Press.

Alexander, J.S., R.C. Wilson, and W.R. Green. 2012. *A Brief History and Summary of the Effects of River Engineering and Dams on the Mississippi River System and Delta.* Circular 1375. U.S. Geological Survey.

Alexander, M. 2013. *Local communities and stakeholders.* In *Management Planning for Nature Conservation* (pp. 41-51). Dordrecht: Springer.

Allison, M.A., G.C. Kineke, E.S. Gordon, and M.A. Goni. 2000. Development and reworking of a seasonal flood deposit on the inner continental shelf off the Atchafalaya River. *Continental Shelf Research* 20:2267-2294.

Allison, M.A., C.R. Demas, B.A. Ebersole, B.A. Kleiss, C.D. Little, E.A. Meselhe, N.J. Powell, T.C. Pratt, and B.M. Vosburg. 2012. A water and sediment budget for the lower Mississippi–Atchafalaya River in flood years 2008–2010: Implications for sediment discharge to the oceans and coastal restoration in Louisiana. *Journal of Hydrology* 432:84-97.

Allison, M., B. Yuill, T. Törnqvist, F. Amelung, T. Dixon, G. Erkens, R. Stuurman, C. Jones, G. Milne, M. Steckler, and J. Syvitski. 2016. Global risks and research priorities for coastal subsidence. *Eos* 97(19):22-27.

Alvarez, L. 2017. A great migration from Puerto Rico is set to transform Orlando. *The New York Times.* November 17, 2017. Available at https://www.nytimes.com/2017/11/17/us/puerto-ricans-orlando.html [accessed March 6, 2018].

Anderson, C.P., G.A. Carter, and W.R. Funderburk. 2016. The use of aerial R, G, B imagery and LIDAR in comparing ecological habitats and geomorphic features on a natural versus man-made barrier island. *Remote Sensing* 8(7):602. doi:10.3390/rs8070602.

Anderson, J.B. 2007. *The formation and future of the upper Texas coast: A geologist answers questions about sand, storms, and living by the sea.* College Station: Texas A&M University Press.

Anderson, J.B., and A.B. Rodriguez (eds.). 2008. *Response of upper gulf coast estuaries to Holocene climate change and sea level rise.* Boulder, CO: The Geological Society of America, Inc.

Anderson, J.B., D.J. Wallace, A.R. Simms, A.B. Rodriguez, and K.T. Milliken. 2014. Variable response of coastal environments of the northwestern Gulf of Mexico to sea level rise and climate change: Implications for future change. *Marine Geology* 352:348-366. doi:10.1016/j.margeo.2013.12.008.

Antolínez, J.A.A., F.J. Méndez, P. Camus, S. Vitousek, E.M. González, P. Ruggiero, and P. Barnard. 2016. A multiscale climate emulator for long-term morphodynamics (MUSCLE-morpho). *Journal of Geophysical Research: Oceans* 121(1):775-791. doi:10.1002/2015JC011107.

Anton, A., J. Cebrian, K. Heck, C.M. Duarte, K.L. Sheehan, M. Miller, and C.D. Foster. 2011. Decoupled effects (positive to negative) of nutrient enrichment on ecosystem services. *Ecological Applications* 21:991-1009.

API (American Petroleum Institute). 2017. *2017 State of American Energy Address.* American Petroleum Institute (API), delivered by Jack Gerard. Available at http://www.api.org/news-policy-and-issues/testimony-and-speeches/2017/01/04/2017-state-of-american-energy-address [accessed March 6, 2018].

Appendini, C.M., A. Torres-Freyermuth, P. Salles, J. López-González, and E. Tonatiuh Mendoza. 2014. Wave climate and trends for the Gulf of Mexico: A 30-year wave hindcast. *Journal of Climate* 27:4. doi: 10.1175/JCLI-D-13-00206.1.

Ariely, D. 2009. *Predictably Irrational.* New York: HarperCollins.

Armitage, A.R., W.E. Highfield, S.D. Brody, and P. Louchouarn. 2015. The contribution of mangrove expansion to salt marsh loss on the Texas Gulf Coast. *PLOS ONE* 10(5):e0125404. https://doi.org/10.1371/journal.pone.0125404.

ASCE (American Society of Civil Engineers). 2003. *Advances in coastal structure design.* Mohan, R.K., O. Magoon, and M. Pirrello (eds.). doi:10.1061/9780784406892.

ASCE. 2017. 2017 *Infrastructure report card.* American Society of Civil Engineers. Available at https://www.infrastructure reportcard.org [accessed April 30, 2018].

Ashton, A.D., and A.B. Murray. 2016. High-angle wave instability and emergent coastal shapes: 1. Modeling of sand waves, flying spits and capes. *Journal of Geophysical Research: Earth Surface* 111(F4). doi:10.1029/2005JF00042.

Atkinson, J., J. McKee Smith, and C. Bender. 2012. Sea-level rise effects on storm surge and nearshore waves on the Texas coast: Influence of landscape and storm characteristics. *Journal of Waterway, Port, Coastal, and Ocean Engineering* 139(2):98-117.

Bankes, S. 1993. Exploratory modeling for policy analysis. *Operations Research* 41(3):435.

Barkwith, A., C.W. Thomas, P.W. Limber, M.A. Ellis, and A.B. Murray. 2014. Assessing the vulnerability of a pinned, soft-cliff coastline—Part I: Assessing the natural sensitivity to wave climate. *Earth Surface Dynamics* 2:295-308. doi:10.5194/esurf-2-295-2014.

BARR. 2014. *Endangered Species Act biological evaluation: In support of West Refinery domestic crude project permit application.* Corpus Christi, TX: Flint Hills Resources Corpus Christi, LLC. Available at https://archive.epa.gov/region6/6pd/air/pd-r/ghg/web/pdf/flinthills-biological-assessment.pdf [accessed March 20, 2018].

Bartunek, J.M., and S.L. Rynes. 2014. Academics and practitioners are alike and unlike. *Journal of Management* 40(5):1181-1201.

Bartunek, J.M., S.L. Rynes, and R.L. Daft. 2001. Across the great divide: Knowledge creation and transfer between practitioners and academics. *Academy of Management Journal* 44(2):340-355.

Bayer, P., N. Keohane, and C. Timmins. 2009. Migration and hedonic valuation: The case of air quality. *Journal of Environmental Economics and Management* 58(1):1-14.

Beavers, J.A., and N.G. Thompson. 2006. External corrosion of oil and natural gas pipelines. *ASM Handbook* 13:1015-1025.

Beck, M.W., K.L. Heck, K.W. Able, D.L. Childers, D.B. Eggleston, B.M. Gillanders, B. Halpern, C.G. Hays, K. Hoshino, T.J. Minello, R.J. Orth, P.F. Sheridan, and M.P. Weinstein. 2001. The identification, conservation, and management of estuarine and marine nurseries for fish and invertebrates. *Bioscience* 51(8):633-641.

Beck, M.W., R.D. Brumbaugh, L. Airoldi, A. Carranza, L.D. Coen, C. Crawford, O. Defeo, G.J. Edgar, B. Hancock, M.C. Kay, H.S. Lenihan, M.W. Luckenbach, C.L. Toropova, G. Zhangand, and X. Guo. 2011. Oyster reefs at risk and recommendations for conservation, restoration, and management. *BioScience* 61:107-116.

Beierle, T.C. 2002. The quality of stakeholder-based decisions. *Risk Analysis* 22(4):739-749.

Ben-Haim, Y. 2001. *Information-gap decision theory: Decisions under severe uncertainty.* San Diego, CA: Academic Press.

Bender, M.A., T.R. Knutson, R.E. Tuleya, J.J. Sirutis, G.A. Vecchi, S.T. Garner, and I.M. Held. 2010. Modeled impact of anthropogenic warming on the frequency of intense Atlantic hurricanes. *Science* 327(5964):454-458.

Bernier, C., J.R. Elliott, J.E. Padgett, F. Kellerman, and P.B. Bedient. 2017. Evolution of social vulnerability and risks of chemical spills during storm surge along the Houston ship channel. *Natural Hazards Review* 18(4). doi:10.1061/(ASCE)NH.1527-6996.0000252.

Bianchi, T.S., J.R. Pennock, and R.R. Twilley (eds.). 1998. *Biogeochemistry of Gulf of Mexico estuaries.* New York: John Wiley & Sons.

Biasutti, M., A.H. Sobel, S.J. Camargo, and T.T. Creyts. 2012. Projected changes in the physical climate of the Gulf Coast and Caribbean. *Climatic Change* 112(3-4):819-845.

Bilskie, M.V., S.C. Hagen, S.C. Medeiros, and D.L. Passeri. 2014. Dynamics of sea level rise and coastal flooding on a changing landscape. *Geophysical Research Letters* 41(3):927-934.

Blum, M.D., and H.H. Roberts. 2009. Drowning of the Mississippi Delta due to insufficient sediment supply and global sea level rise. *Nature Geoscience* 2(7):488-491. doi:10.1038/ngeo553.

BOEM (Bureau of Ocean Energy Management). Austin, D., and D. Woodson (eds.). 2014. *Gulf Coast communities and the fabrication and shipbuilding industry: A comparative community study, Volume II: Community Profiles*. New Orleans, LA: Gulf of Mexico OCS Region. OCS Study BOEM 2014-610.

Bonney, R., C.B. Cooper, J. Dickinson, S. Kelling, T. Phillips, K.V. Rosenberg, and J. Shirk. 2009. Citizen science: A developing tool for expanding science knowledge and scientific literacy. *BioScience* 59(11):977-984.

Bostick, T.P., T.H. Holzer, and S. Sarkani. 2017. Enabling stakeholder involvement in coastal disaster resilience planning. *Risk Analysis* 37(6):1181-1200. doi:10.1111/risa.12737.

Bradbury, J., M. Allen, and R. Dell. 2015. Climate change and energy infrastructure exposure to storm surge and sea level rise. Available at https://www.energy.gov/sites/prod/files/2015/07/f24/QER%20Analysis%20-%20Climate%20Change%20and%20Energy%20Infrastructure%20Exposure%20to%20Storm%20Surge%20and%20Sea-Level%20Rise_0.pdf [accessed May 15, 2018].

Brashares, J.S., B. Abrahms, K.J. Fiorella, C.D. Golden, C.E. Hojnowski, R.A. Marsh, D.J. McCauley, T.A. Nuñez, K. Seto, and L. Withey. 2014. Wildlife decline and social conflict. *Science* 345(6195):376-378.

Breton, S.P., and G. Moe. 2009. Status, plans and technologies for offshore wind turbines in Europe and North America. *Renewable Energy* 34:646-654.

Brown, S., B. Couvillion, K. de Mutsert, J. Fischbach, H. Roberts, M. Rodrigue, J. Schindler, G. Thomson, J. Visser, and E. White. 2017. *2017 Coastal Master Plan: Appendix C: Modeling Chapter 3—Modeling Components and Overview. Version final.* (p. 72). Baton Rouge, LA: Coastal Protection and Restoration Authority.

Bruine de Bruin, W., and A. Bostrom. 2013. Assessing what to address in science communication. *Proceedings of the National Academy of Sciences of the United States of America* 110(Supplement 3):14062-14068.

Bruun, P. 1972. *The history and philosophy of coastal protection.* 13th International Conference on Coastal Engineering, July 10–14, 1972. Vancouver, British Columbia, Canada. doi:10.1061/9780872620490.003

Bryant, B.P., and R.J. Lempert. 2010. Thinking inside the box: A participatory, computer-assisted approach to scenario discovery. *Technological Forecasting and Social Change* 77(1):34-49. doi: 10.1016/j.techfore.2009.08.002.

Bubeck, P., W.J. Botzen, and J.C. Aerts. 2012. A review of risk perceptions and other factors that influence flood mitigation behavior. *Risk Analysis* 32(9):1481-1495. doi:10.1111/j.1539-6924.2011.01783.x.

Bukulich, A. 2013. Explanatory models versus predictive models: Reduced complexity modeling in geomorphology. In V. Karakostas and D. Dieks (eds.). EPSA11 Perspectives and Foundational Problems in Philosophy of Science. *The European Philosophy of Science Association Proceedings 2*. doi:10.1007/978-3-319-01306-0__10.

Bunten, D., and M.E. Kahn. 2017. Optimal real estate capital durability and localized climate change disaster risk. *Journal of Housing Economics* 36:1-7.

Burrows, M.T., D.S. Schoeman, L.B. Buckley, P. Moore, E.S. Poloczanska, K.M. Brander, C. Brown, J.F. Bruno, C.M. Duarte, B.S. Halpern, and J. Holding. 2011. The pace of shifting climate in marine and terrestrial ecosystems. *Science* 334(6056):652-655.

Byrnes, M.R., and J.L. Berlinghoff. 2012. Gulf regional sediment management master plan: Case study compilation. *Journal of Coastal Research* 60:72-124.

Camus, P., F.J. Méndez, I.J. Losada, M. Menéndez, A. Espejo, J. Pérez, A. Rueda, and Y. Guanche. 2014a. A method for finding the optimal predictor indices for local wave climate conditions. *Ocean Dynamics* 64(7):1025-1038. doi:10.1007/s10236-014-0737-2.

Camus, P., M. Menéndez, F.J. Méndez, C. Izaguirre, A. Espejo, V. Cánovas, J. Pérez, A. Rueda, I.J. Losada, and R. Medina. 2014b. A weather-type statistical downscaling framework for ocean wave climate. *Journal of Geophysical Research: Oceans* 119(11):7389-7405. doi:10.1002/2014JC010141.

Carpenter, S.R., H.A. Mooney, J. Agard, D. Capistrano, R.S. DeFries, S. Díaz, T. Dietz, A.K. Duraiappah, A. Oteng-Yeboah, H.M. Pereira, and C. Perrings. 2009. Science for managing ecosystem services: Beyond the Millennium Ecosystem Assessment. *Proceedings of the National Academy of Sciences of the United States of America* 106(5):1305-1312.

CCSP (U.S. Climate Change Science Program). 2007. *Effects of climate change on energy production and use in the United States*. A report by the U.S. Climate Change Science Program and the subcommittee on Global Change Research. T.J. Wilbanks, V. Bhatt, D.E. Bilello, S.R. Bull, J. Ekmann, W.C. Horak, Y.J. Huang, M.D. Levine, M.J. Sale, D.K. Schmalzer, and M.J. Scott (eds.). Washington, DC: Department of Energy, Office of Biological and Environmental Research.

CCSP. 2008. *Impacts of climate change and variability on transportation systems and infrastructure: Gulf Coast study*. U.S. Climate Change Science Program (CCSP) Synthesis and Assessment Product 4.7. March 2008.

Chang, C., E. Mallman, and M. Zoback. 2014. Time-dependent subsidence associated with drainage-induced compaction in Gulf of Mexico shales bounding a severely depleted gas reservoir. *AAPG Bulletin* 98(6):1145-1159.

Charlier, R.H., M.C.P. Chaineux, and S. Morcos. 2005. Panorama of the history of coastal protection. *Journal of Coastal Research* 21(1):79-111. doi:10.2112/03561.1.

Chen, Y., J. Cebrian, J.C. Lehrter, B. Christiaen, J. Stutes, and J. Goff. 2017. Storms do not alter long-term watershed development influences on coastal water quality. *Marine Pollution Bulletin* 122(1-2):207-216. doi:10.1016/j.marpolbul.2017.06.038.

Chesapeake Bay Program. 1983. *The Chesapeake Bay Agreement of 1983*. Available at https://www.chesapeakebay.net/documents/1983_CB_Agreement2.pdf [accessed May 15, 2018].

Chesapeake Bay Program. 2014. *Chesapeake Bay Watershed Agreement*. Available at https://www.chesapeakebay.net/documents/FINAL_Ches_Bay_Watershed_Agreement.withsignatures-HIres.pdf [accessed May 15, 2018].

Chesapeake Bay Program. 2018a. *Modeling the Chesapeake Bay Watershed*. Available at https://www.chesapeakebay.net/what/programs/modeling [accessed February 21, 2018].

Chesapeake Bay Program. 2018b. *Monitoring*. Available at https://www.chesapeakebay.net/what/programs/monitoring [accessed February 21, 2018].

Christiaen, B., A. McDonald, J. Cebrian, and A. Ortmann. 2013. Response of the microbial community to environmental change during seagrass transplantation. *Aquatic Botany* 109:31-38.

Christiaen, B., J.C. Lehrter, J. Goff, and J. Cebrian. 2016. Functional implications of changes in seagrass species composition in two shallow coastal lagoons. *Marine Ecology Progress Series* 557:111-121.

Cloern, J.E. 2001. Our evolving conceptual model of the coastal eutrophication problem. *Marine Ecology Progress Series* 210:223-253.

Coleman, F.C., C.C. Koenig, G.R. Huntsman, J.A. Musick, A.M. Eklund, J.C. McGovern, R.W. Chapman, G.R. Sedberry, and C.B. Grimes. 2000. Long-lived reef fishes: The grouper-snapper complex. *Fisheries* 25(3):14-21.

Coleman, F.C., W.F. Figueira, J.S. Ueland, and L.B. Crowder. 2004. The impact of United States recreational fisheries on marine fish populations. *Science* 305(5692):1958-1960.

Colten, C.E. 2005. *An unnatural metropolis: Wresting New Orleans from nature*. Baton Rouge: Louisiana State University Press.

Colten, C.E., R.W. Kates, and S.B. Laska. 2008. Three years after Katrina: Lessons for community resilience. *Environment: Science and Policy for Sustainable Development* 50(5):36-47. doi:10.3200/ENVT.50.5.36-47.

Consortium for Resilient Gulf Communities. 2018. *About the Consortium for Resilient Gulf Communities*. Available at https://www.resilientgulf.org/about-us/ [accessed April 24, 2018].

Cowan, J.H., C.B. Grimes, W.F. Patterson III, C.J. Walters, A.C. Jones, W.J. Lindberg, D.J. Sheehy, W.E. Pine III, J.E. Powers, M.D. Campbell, K.C. Lindeman, S.L. Diamond, R. Hilborn, H.T. Gibson, and A. Rose. 2011. Red snapper management in the Gulf of Mexico: Science- or faith-based? *Reviews in Fish Biology and Fisheries* 21(2):187-204.

Cowell, P.J., P.S. Roy, and R.A. Jones. 1995. Simulation of large-scale coastal change using a morphological behaviour model. *Marine Geology* 126(1-4):45-61. doi:10.1016/0025-3227(95)00065-7.

Cranswick, D. 2001. *Brief overview of Gulf of Mexico OCS oil and gas pipelines: Installation, potential impacts, and mitigation measures*. New Orleans, LA: U.S. Department of the Interior, Minerals Management Service, Gulf of Mexico OCS Region. OCS Report MMS 2001-067.

Culver, S.J., C.A. Grand Pre, D.J. Mallinson, S.R. Riggs, D.R. Corbett, J. Foley, M. Hale, L. Metger, J. Ricardo, J. Rosenberger, C.G. Smith, C.W. Smith, S.W. Snyder, D. Twmaley, K. Farrell, and B. Horton. 2007. Late Holocene barrier island collapse: Outer Banks, North Carolina, USA. *The Sedimentary Record* 5(4):4-8.

Cunniff, S., and A. Schwartz. 2015. *Performance of natural infrastructure and nature-based measures as coastal risk reduction features*. Washington, DC: Environmental Defense Fund.

Cutter, S.L., and C.T. Emrich. 2006. Moral hazard, social catastrophe: The changing face of vulnerability along the hurricane coasts. *The Annals of the American Academy of Political and Social Science* 604(1):102-112.

Cutter, S.L., C.T. Emrich, J.T. Mitchell, B.J. Boruff, M. Gall, M.C. Schmidtlein, C.G. Burton, and G. Melton. 2006. The long road home: Race, class, and recovery from Hurricane Katrina. *Environment: Science and Policy for Sustainable Development* 48(2):8-20. doi: 10.3200/ENVT.48.2.8-20.

Cutter, S.L., L. Barnes, M. Berry, C. Burton, E. Evans, E. Tate, J. Webb. 2008. A place-based model for understanding community resilience to natural disasters. *Global Environmental Change* 18:598-606.

Dame, R.F. 1993. *The role of bivalve filter feeder material fluxes in estuarine ecosystems* (pp. 245-269). In *Bivalve filter feeders in estuarine and coastal ecosystem processes*. Heidelberg: Springer.

Day, J.W., R.R. Christian, D.M. Boesch, A. Yáñez-Arancibia, J. Morris, R.R. Twilley, L. Naylor, L. Schaffner, and C. Stevenson. 2008. Consequences of climate change on the ecogeomorphology of coastal wetlands. *Estuaries and Coasts* 31(3):477-491.

Day, J.W., D.F. Boesch, E.J. Clairain, G.P. Kemp, S.B. Laska, W.J. Mitsch, K. Orth, H. Mashriqui, D.J. Reed. L. Shabman, and C.A. Simenstad. 2017. Restoration of the Mississippi Delta: Lessons from Hurricanes Katrina and Rita. *Science* 315(5819):1679-1684.

de Groot, R.S., R. Alkemade, L. Braat, L. Hein, and L. Willemen. 2010. Challenges in integrating the concept of ecosystem services and values in landscape planning, management, and decision-making. *Ecological Complexity* 7:260-272.

De Nooy, W. 2013. Communication in natural resource management: Agreement between and disagreement within stakeholder groups. *Ecology and Society* 18(2).

de Vriend, H.J., M. van Koningsveld, S.G. Aarninkhof, M.B. de Vries, and M.J. Baptist. 2015. Sustainable hydraulic engineering through building with nature. *Journal of Hydro-Environment Research* 9(2):159-171.

Dean, R.G., and R.A. Dalrymple. 2002. *Coastal processes with engineering applications*. Cambridge: Cambridge University Press.

DeConto, R.M., and D. Pollard. 2016. Contribution of Antarctica to past and future sea level rise. *Nature* 531(7596):591-597.

DeLaune, R.D., J.A. Nyman, and W.H. Patrick. 1994. Peat collapse, ponding, and wetland loss in a submerging coastal marsh. *Journal of Coastal Research* 10(4):1021-1030.

DeLorme, D.E., D. Kidwell, S.C. Hagen, and S.H. Stephens. 2016. Developing and managing transdisciplinary and transformative research on the coastal dynamics of sea level rise: Experiences and lessons learned. *Earth's Future* 4(5):194-209.

DHS (U.S. Department of Homeland Security). 2010. Aging infrastructure: Issues, research, and technology. *Buildings and Infrastructure Protection Series*, BIPS 01.

DHS-DOE (U.S. Department of Homeland Security-U.S. Department of Energy). 2015. *Energy sector-specific plan 2015*. Washington, DC: DHS, DOE.

Dietz, T. 2013. Bringing values and deliberation to science communication. *Proceedings of the National Academy of Sciences of the United States of America* 110(Supplement 3):14081-14087.

Dilling, L., and M.C. Lemos. 2011. Creating usable science: Opportunities and constraints for climate knowledge use and their implications for science policy. *Global Environmental Change* 21(2):680-689.

Dismukes, D.E. 2011. *OCS-related infrastructure factbook, Vol. 1, Post-hurricane impact assessment*, OCS Study BOEM 2011-043. U.S. Department of Interior, Bureau of Ocean Energy Management (BOEM), Gulf of Mexico OCS Region.

Dixon, T.H., F. Amelung, A. Ferretti, F. Novali, F. Rocca, F., R. Dokka, G. Sellell, S-W Kim, S. Wdowinski, and D. Whitman. 2006. Space geodesy: Subsidence and flooding in New Orleans. *Nature* 441:587-588. doi:10.1038/441587a.

DOE (U.S. Department of Energy). 2010. *Hardening and resiliency, U.S. energy industry response to recent hurricane seasons, Infrastructure Security and Energy Restoration*. U.S. Department of Energy. August 2010.

DOE. 2015. *Climate change and the U.S. energy sector: Regional vulnerabilities and resilience solutions*. Available at https://www.energy.gov/sites/prod/files/2015/10/f27/Regional_Climate_Vulnerabilities_and_Resilience_Solutions_0.pdf [accessed May 3, 2018].

DOE. 2016. *Long-term strategic review of the U.S. Strategic Petroleum Reserve*. Report to Congress.

DOT (U.S. Department of Transportation). 2011. *The state of the national pipeline infrastructure.* Pipeline and Hazardous Materials Safety Administration, U.S. Department of Transportation.

Douglas, B.C. 1991. Global sea level rise. *Journal of Geophysical Research* 96(C4):6981-6992.

Doyle, T.W. 2009. Hurricane frequency and landfall distribution for coastal wetlands of the Gulf Coast, USA. *Wetlands* 29(1):35-43.

Draut, A.E., G.C. Kineke, O.K. Huh, J.M. Grymes III, K.A. Westphal, and C.C. Moeller. 2005. Coastal mudflat accretion under energetic conditions, Louisiana Chenier-Plain Coast, USA. *Marine Geology* 214(1-3):27-47.

Duchmann, H. 2018. *Port Fourchon expansion moving forward.* Available at http://www.houmatoday.com/news/20180325/port-fourchon-expansion-moving-forward [accessed May 10, 2018].

Dunn, D.D. 1996. *Trends in nutrient inflows to the Gulf of Mexico from streams draining the conterminous United States, 1972–93.* 96-4113. USGS Water-Resources Investigation.

Durán, O., and L.J. Moore. 2013. Vegetation controls on the maximum size of coastal dunes. *Proceedings of the National Academy of Sciences of the United States of America* 110(43):17217-17222.

Durán, O., and L.J. Moore. 2015. Barrier island bistability induced by biophysical interactions. *Nature Climate Change* 5(2):158-162.

Dzwonkowski, B., K. Park, and L. Jiang. 2011. Subtidal across-shelf velocity structure and surface transport effectiveness on the Alabama shelf of the northeastern Gulf of Mexico. *Journal of Geophysical Research: Oceans* 116(C10).

Egan, M. 2017. *Gas shortage: Florida races to refill gas stations after Hurricane Irma.* Available at http://gmoney.cnn.com/2017/09/11/investing/gas-station-shortage-florida-hurricane-irma/index.html [accessed May 2, 2018].

Ehrenfeld, J.G. 1990. Dynamics and processes of barrier island vegetation. *Reviews in Aquatic Sciences* 2(3-4):437-480.

EIA (U.S. Energy Information Administration). 2017a. U.S. energy mapping system, U.S. Energy Information Administration. Available at https://www.eia.gov/state/maps.php?v=Petroleum [accessed May 23, 2017].

EIA. 2017b. Number and capacity of operable petroleum refineries by PAD district and state as of January 1, 2017. U.S. Energy Information Administration. Available at https://www.eia.gov/petroleum/refinerycapacity/table1.pdf [accessed March 15, 2018].

EIA. 2017c. Gulf of Mexico fact sheet. U.S. Energy Information Administration. Available at https://www.eia.gov/special/gulf_of_mexico/ [accessed March 15, 2018].

EIA. 2018. Annual energy outlook 2018, with projections to 2050. Available at http://pinnaclepetroleum.com/Press-News/eias-annual-energy-outlook-2018-with-projections-to-2050 [accessed May 15, 2018].

Eleuterius, C.K., and S.L. Beaugez. 1979. *Mississippi Sound: A hydrographic and climatic atlas. Physical Oceanography Section.* Mississippi-Alabama Sea Grant Consortium. Gulf Coast Research Laboratory, Ocean Springs, Mississippi. MASGP-79-009.

Elko, N., F. Feddersen, D. Foster, C. Hapke, J. McNinch, R. Mulligan, H.T. Özkan-Haller, N. Plant, and B. Raubenheimer. 2014. *The future of nearshore processes research.* The Nearshore Processes Community.

Emery, S.M., L. Bell-Dereske, and J.A. Rudgers. 2015. Fungal symbiosis and precipitation alter traits and dune building by the ecosystem engineer, *Ammophila breviligulata. Ecology* 96(4):927-935. doi:10.1890/14-1121.1.

EPA (U.S. Environmental Protection Agency). 1999. *The ecological conditions of estuaries in the Gulf of Mexico.* Available at https://www.epa.gov/gulfofmexico/ecological-condition-estuaries-gulf-mexico [accessed May 15, 2018].

EPA. 2017. *Public participation guide: Introduction to public participation.* Available at https://www.epa.gov/international-cooperation/public-participation-guide-introduction-public-participation [accessed May 15, 2018].

Esposito, C.R., Y.G. Ioannis, and A.S. Kolker. 2013. Hydrodynamic and geomorphic controls on mouth bar evolution. *Geophysical Research Letters* 40(8):1540-1545.

Esposito, C.R., Z. Shen, T.E. Törnqvist, J. Marshak, and C. White. 2017. Efficient retention of mud drives land building on the Mississippi Delta plain. *Earth Surface Dynamics* 5(3):387-397. doi:10.5194/esurf-5-387-2017.

Feagin, R.A., D.J. Sherman, and W.E. Grant. 2005. Coastal erosion, global sea-level rise, and the loss of sand dune plant habitats. *Frontiers in Ecology and the Environment* 3(7):359-364.

Feagin, R.A., W.K. Smith, N.P. Psuty, D.R. Young, M.L. Martinez, G.A. Carter, K.L. Lucas, J.C. Gibeaut, J.N. Gemma, and R.E. Koske. 2010. Barrier islands: Coupling anthropogenic stability with ecological sustainability. *Journal of Coastal Research* 26(6):987-992.

Feagin, R.A., K.M. Yeager, C.A. Brunner, and J.G. Paine. 2013. Active fault motion in a coastal wetland: Matagorda, Texas. *Geomorphology* 199:150-159. doi:10.1016/j.geomorph.2012.08.013.

Feagin, R.A., J. Figlus, J.C. Zinnert, J. Sigren, M.L. Martinez, R. Silva, W.K. Smith, D. Cox, D.R. Young, and G. Carter. 2015. Going with the flow or against the grain? The promise of vegetation for protecting beaches, dunes and barrier islands from erosion. *Frontiers in Ecology and the Environment* 13(4):203-210.

Fearnley, S., M. Miner, M. Kulp, C. Bohling, and S. Penland. 2009. Hurricane impact and recovery shoreline change analysis of the Chandeleur Islands, Louisiana, USA: 1855 to 2005. *Geo-Marine Letters* 29(6):455–466.

Feldman, D.L., and H.M. Ingram. 2009. Making science useful to decision makers: Climate forecasts, water management, and knowledge networks. Weather, Climate, and Society 1(1):9-21.

FEMA (Federal Emergency Management Agency). 2009. *FEMA P-757, Hurricane Ike in Texas and Louisiana, mitigation team assessment report, building performance observations, recommendations and technical guidance.* Federal Emergency Management Agency.

FEMA. 2018. Data visualization: Disaster declarations for states and counties. Available at https://www.fema.gov/data-visualization-disaster-declarations-states-and-counties [accessed May 2, 2018].

Fischbach, J.R., D.R. Johnson, D.S. Ortiz, B.P. Bryant, M. Hoover, and J. Ostwald. 2012. *Coastal Louisiana Risk Assessment Model: Technical description and 2012 Coastal Master Plan analysis results.* Santa Monica, CA: RAND Corporation, TR-1259-CPRA.

Fischbach, J.R., D.R. Johnson, and E. Molina-Perez. 2017. *Reducing coastal flood risk with a Lake Pontchartrain barrier.* Santa Monica, CA: RAND Corporation. Available at: https://www.rand.org/pubs/research_reports/RR1988.html [accessed May 2, 2018].

Fischhoff, B. 2013. The sciences of science communication. *Proceedings of the National Academy of Sciences of the United States of America* 110(Supplement 3):14033-14039.

Fischhoff, B., and D.A. Scheufele. 2013. The science of science communication. *Proceedings of the National Academy of Sciences of the United States of America* 110(Supplement 3):14031-14032.

Fitzpatrick, C., A. Kolker, and P. Chu. 2017. *Variation in the Mississippi River plume from data synthesis of model outputs and MODIS imagery*, OS23A-1379. Presented at the 2017 American Geophysical Union Fall Meeting, New Orleans, LA.

Fodrie, F., K. Heck, S.P. Powers, W.M. Graham, and K.L. Robinson. 2010. Climate-related, decadal-scale assemblage changes of seagrass-associated fishes in the northern Gulf of Mexico. *Global Change Biology* 16:48-59.

French, J., A. Payo, A.B. Murray, J. Orford, M. Eliot, and P. Cowell, P. 2016. Appropriate complexity for the prediction of coastal and estuarine geomorphic behavior at decadal to centennial scales. *Geomorphology* 256:3-16. doi:10.1016/j.geomorph.2015.10.005.

Fritz, H.M., C. Blount, R. Sokoloski, J. Singleton, A. Fuggle, B.G. McAdoo, A. Moore, C. Grass, and B. Tate. 2007. Hurricane Katrina storm surge distribution and field observations on the Mississippi barrier islands. *Estuarine, Coastal and Shelf Science* 74(1-2):12-20.

Funderburk, W.R., G.A. Carter, and C.P. Anderson. 2016. Evaluating the influence of elevation and impact of Hurricane Katrina on radial growth in slash pine (*Pinus elliottii* var. *elliottii* Engelm) on Cat Island, Mississippi, USA. *Journal of Coastal Research* 32(3):483-489. doi:10.2112/JCOASTRES-D-15-00038.1.

Galloway, W.E. 2008. *Chapter 15 depositional evolution of the Gulf of Mexico sedimentary basin* (pp. 505-549). In Andrew D. Miall (ed.). *Sedimentary Basins of the World, Vol. 5. The Sedimentary Basins of the United States and Canada.* Amsterdam, Netherlands: Elsevier.

Garrote-Moreno, A., A. McDonald, T.D. Sherman, J.L. Sanchez-Lizaso, K.L. Heck, and J. Cebrian. 2014. Short-term impacts of salinity pulses on ionic ratios of the seagrasses Thalassia testudinum and Halodule wrightii. *Aquatic Botany* 120:315-321.

Gehrels, W.R., and P.L. Woodworth. 2013. When did modern rates of sea-level rise start? *Global and Planetary Change* 100:263-277. doi:10.1016/j.gloplacha.2012.10.020.

Gibeaut, J.C., W.A. White, T. Hepner, R. Gutiérrez, T.A. Tremblay, R. Smyth, and J. Andrews. 2000. *Texas Shoreline Change Project: Gulf of Mexico shoreline change from the Brazos River to Pass Cavallo.* A Report of the Texas Coastal Coordination Council. Contract no. NA870Z0251.

Gibeaut, J.C., W.A. White, R.C. Smyth, J.R. Andrews, T.A. Tremblay, R. Gutiérrez, T.L. Hepner, and A. Neuenschwander. 2003. Topographic variation of barrier island subenvironments and associated habitats. *Proceedings of the Fifth International Symposium on Coastal Engineering and Science of Coastal Sediment Processes* (Clearwater Beach, Florida), Coastal Sediments 2003: Crossing disciplinary boundaries.

Godfrey, P.J., S.P. Leatherman, and R. Zaremba. 1979. *A geobotanical approach to classification of barrier beach systems* (pp. 99-126). In S.P. Leatherman (ed). *Barrier islands from the Gulf of St. Lawrence to the Gulf of Mexico*. New York: Academic Press.

Goldthau, A. 2014. Rethinking the governance of energy infrastructure: Scale, decentralization, and polycentrism. *Energy Research and Social Science* 1:134-140. doi:0.1016/j.erss.2014.02.009.

Gopalakrishnan, S., M.D. Smith, J.M. Slott, and A.B. Murray. 2011. The value of disappearing beaches: A hedonic pricing model with endogenous beach width. *Journal of Environmental Economics and Management* 61(3):297-310.

Gopalakrishnan, S., C. Landry, M.D. Smith, and J.C. Whitehead. 2016. Economics of coastal erosion and adaptation to sea level rise. *Annual Review of Resource Economics* 8:13.1-13.21.

Gordon, R.J. 2016. *The rise and fall of American growth: The U.S. standard of living since the Civil War*. Princeton, NJ: Princeton University Press.

Granek, E.F., S. Polasky, C.V. Kappel, D.J. Reed, D.M. Stoms, E.W. Koch, C.J. Kennedy, L.A. Cramer, S.D. Hacker, E.B. Barbier, S. Aswani, M. Ruckelshaus, G.M.E. Perillo, B.R. Silliman, N. Muthiga, D. Bael, and E. Wolanski. 2010. Ecosystem services as a common language for coastal ecosystem-based management. *Conservation Biology* 24(1):207-216.

Grice, P. 1975. *Logic and conversation*. In P. Cole and J. Morgan (eds.). *Syntax and semantics. 3: Speech acts*. New York: Academic Press. pp. 41-58.

Groves, D.G., and R.J. Lempert. 2007. A new analytic method for finding policy-relevant scenarios. *Global Environmental Change Part A: Human & Policy Dimensions* 17(1):73-85. doi: 10.1016/j.gloenvcha.2006.11.006.

Groves, D.G., J.R. Fischbach, E. Bloom, D. Knopman, and R. Keefe. 2013. *Adapting to a changing Colorado River: Making future water deliveries more reliable through robust management strategies*. Santa Monica, CA: RAND Corporation, RR-242-BOR.

Guidry, L. 2016. Port signs agreements with Wisner Trust and Corps of Engineers, plans next generation of port facilities. September 29, 2016. Available at http://portfourchon.com/port-signs-agreements [accessed May 10, 2018].

Guidry, L. 2018. GLPC and Corps sign assumption of management agreement. April 5, 2018. Available at http://portfourchon.com/glpc-and-corps-sign-assumption-of-maintenance-agreement [accessed May 10, 2018].

Guston, D.H., W. Clark, T. Keating, D. Cash, S. Moser, C. Miller, and C. Powers. 2000. *Report of the workshop on boundary organizations in environmental policy and science*. B 9-10 December 1999, Bloustein School of Planning and Public Policy, Rutgers University, New Brunswick, NJ. Belfer Center for Science and International Affairs (BCSIA) Discussion Paper 2000-32. Piscataway, NJ: Environmental and Occupational Health Sciences Institute at Rutgers University and UMDNJ-RWJMS; Cambridge, MA: Global Environmental Assessment Project, Environment and Natural Resources Program, Kennedy School of Government, Harvard University.

Gutmann, M.P., and V. Field. 2010. Katrina in historical context: Environment and migration in the U.S. *Population and Environment* 31(1-3):3-19.

Haasnoot, M., J.H. Kwakkel, W.E. Walker, and J. Maat. 2013. Dynamic adaptive policy pathways: A method for crafting robust decisions for a deeply uncertain world. *Global Environmental Change* 23(2):485-498. doi:10.1016/j.gloenvcha.2012.12.006.

Hacker, S., P. Zarnetske, and E. Seabloom. 2012. Subtle differences in two non-native congeneric beach grasses significantly affect their colonization, spread, and impact. *Oikos* 121(1):138-148.

Hall, J.A., S. Gill, J. Obeysekera, W. Sweet, K. Knuuti, and J. Marburger. 2016. *Regional sea level scenarios for coastal risk management: Managing the uncertainty of future sea level change and extreme water levels for Department of Defense coastal sites worldwide*. U.S. Department of Defense, Strategic Environmental Research and Development Program.

Halligan, J.J., M.R. Waters, A. Perrotti, I.J. Owens, J.M. Feinberg, M.D. Bourne, B. Fenerty, B. Winsborough, D. Carlson, D.C. Fisher, T.W. Stafford, Jr., and J.S. Dunbar. 2016. Pre-Clovis occupation 14,550 years ago at the Page-Ladson Site, Florida, and the peopling of the Americas. *Science Advances* 2(5):e1600375. doi:10.1126/sciadv.1600375.

Halverson, J.B. 2018. The costliest hurricane season in U.S. history. *Weatherwise* 71(2):20-27.

Handley, L., D. Altsman, and R. DeMay (eds.). 2007. *Seagrass status and trends in the northern Gulf of Mexico: 1940–2002*. Scientific Investigations Report 2006-5287. U.S. Environmental Protection Agency 855-R-04-003.

Hansen, B. 2007. Weathering the storm: The Galveston seawall and grade raising. *Civil Engineering Magazine Archive* 77(4):32-33.

Hanson, H., 1989. GENESIS: A generalized shoreline change numerical model. *Journal of Coastal Research* (pp. 1-27).

Harrington, M. 2017. *For the first time in 17 years, people's trust declined in every kind of institution we asked about*. Available at https://hbr.org/2017/01/survey-peoples-trust-has-declined-in-business-media-government-and-ngos [accessed May 10, 2018].

Harris, S.P., and Wilson, D.O. 2008. Mitigating hurricane storm surge perils at the DeLisle plant. *Process Safety Progress* 27(3):177-184.

Harthill, R. 2014. Anatomy of a groin. *Your Observer*. Available at https://www.yourobserver.com/article/anatomy-groin [accessed May 10, 2018].

Hassol, S.J., 2008. Improving how scientists communicate about climate change. *Eos* 89(11):106-107.

Hauer, M. 2017. Migration induced by sea-level rise could reshape the U.S. population landscape. *Nature Climate Change* 7:321-325.

Hauer, M.E., J.M. Evans, and D.R. Mishra. 2016. Millions projected to be at risk from sea-level rise in the continental United States. *Nature Climate Change* 6(7):691-695. doi:10.1038/nclimate2961.

Hauxwell, J., J. Cebrian, and I. Valiela. 2003. Eelgrass *Zostera marina* loss in temperate estuaries: Relationship to land-derived nitrogen loads and effect of light limitation imposed by algae. *Marine Ecology Progress Series* 247:59-73.

Hayden, B.P., M.C.F.V., Santos, G. Shao, and R.C. Kochel. 1995. Geomorphological controls on coastal vegetation at the Virginia Coast Reserve. *Biogeomorphology, Terrestrial and Freshwater Systems* 13:283-300.

Hendricks, M.D., M.A. Meyer, N.G. Gharaibeh, S. Van Zandt, J. Masterson, J.T. Cooper, J.A. Horney, and P. Berke. 2018. The development of a participatory assessment technique for infrastructure: Neighborhood-level monitoring towards sustainable infrastructure systems. *Sustainable Cities and Society* 38:265-274.

Herman, J., P. Reed, H. Zeff, and G. Characklis. 2015. How should robustness be defined for water systems planning under change? *Journal of Water Resources Planning and Management* 141(10) doi:10.1061/(ASCE)WR.1943-5452.0000509.

Herrera-Silveira, J.A., and S.M. Morales-Ojeda. 2009. Evaluation of the health status of a coastal ecosystem in southeast Mexico: Assessment of water quality, phytoplankton, and submerged aquatic vegetation. *Marine Pollution Bulletin* 59(1-3):72-86.

Hickey, T.D., A.C. Hine, E.A. Shinn, S.E. Kruse, and R.Z. Poore. 2010. Pleistocene carbonate stratigraphy of south Florida: Evidence for high-frequency sea-level cyclicity. *Journal of Coastal Research* 26(4):605-614.

Hieb, E.E., R.H. Carmichael, A. Aven, C. Nelson-Seely, and N. Taylor. 2017. Sighting demographics of the West Indian manatee *Trichechus manatus* in the north-central Gulf of Mexico supported by citizen-sourced data. *Endangered Species Research* 32:321-332.

Hochschild, A. 2016. *Strangers in their own land: Anger and mourning on the American right*. New York: The New Press.

Horan, R.D., E.P. Fenichel, K.L. Drury, and D.M. Lodge. 2011. Managing ecological thresholds in coupled environmental–human systems. *Proceedings of the National Academy of Sciences of the United States of America* 108(18):7333-7338.

Horowitz, A.J., K.A. Elrick, and J.J. Smith. 2001. Estimating suspended sediment and trace element fluxes in large river basins: Methodological considerations as applied to the NASQAN Programme. *Hydrological Processes* 15(7):1107-1132.

Hoteit, I., T. Hoar, G. Gopalakrishnan, N. Collins, J. Anderson, B. Cornuelle, A. Köhl, and P. Heimbach. 2013. A MITgcm/DART ensemble analysis and prediction system with application to the Gulf of Mexico. *Dynamics of Atmospheres and Oceans* 63:1-23.

Houser, C., and S. Hamilton. 2009. Sensitivity of post-hurricane beach and dune recovery to event frequency. *Earth Surface Processes and Landforms* 34(5):613-628. doi:10.1002/esp.1730.

Howes, N.C., D.M. FitzGerald, Z.J. Hughes, I.Y. Georgiou, M.A. Kulp, M.D. Miner, J.M. Smith, and J.A. Barras. 2010. Hurricane-induced failure of low salinity wetlands. *Proceedings of the National Academy of Sciences of the United States of America* 107(32):14014-14019.

Hsee, C.K., and E.U. Weber. 1999. Cross-national differences in risk preference and lay predictions. *Journal of Behavioral Decision Making* 12(2):165-179.

HUD (U.S. Department of Housing and Urban Development). 2017. *The resettlement of Isle de Jean Charles: Report on data gathering and engagement phase.* Prepared by Concordia with Chicago Bridge & Iron Company, directed by Pan American Engineers, for the State of Louisiana Division of Administration, Office of Community Development, Disaster Recovery Unit.

Hunter, A., J. Cebrian, J.P. Stutes, D. Patterson, B. Christiaen, C. Lafabrie, and J. Goff. 2015. Magnitude and trophic fate of black needlerush (*Juncus roemerianus*) productivity: Does nutrient addition matter? *Wetlands* 35:401-417.

Hurst, M.D., A. Barkwith, M.A. Ellis, C.W. Thomas, and A.B. Murray. 2015. Exploring the sensitivities of crenulate bay shorelines to wave climates using a new vector-based one-line model. *Journal of Geophysical Research: Earth Surface* 120(12):2586-608. doi:10.1002/2015JF003704.

Interagency Climate Change Adaptation Task Force. 2011. *Federal actions for a climate resilient nation: Progress report of the Interagency Climate Change Adaptation Task Force.*

Irish, J.L., and D.T. Resio. 2010. A hydrodynamics-based surge scale for hurricanes. *Ocean Engineering* 37(1):69-81. doi:10.1016/j.oceaneng.2009.07.012.

Irish, J.L., D.T. Resio, and J.J. Ratcliff. 2008. The influence of storm size on hurricane surge. *Journal of Physical Oceanography* 38(9):2003-2013.

Irish, J.L., A.E. Frey, J.D. Rosati, F. Olivera, L.M. Dunkin, J.M. Kaihatu, C.M. Ferreira, and B.L. Edge. 2010. Potential implications of global warming and barrier island degradation on future hurricane inundation, property damages, and population impacted. *Ocean & Coastal Management* 53(10):645-657.

Irish, J.L., A. Sleath, M.A. Cialone, T.R. Knutson, and R.E. Jensen. 2014. Simulations of Hurricane Katrina (2005) under sea level and climate conditions for 1900. *Climatic Change* 122(4):635-649.

Jackson, J.B.C., M.X. Kirby, W.H. Berger, K.A. Bjorndal, L.W. Botsford, B.J. Bourque, R.H. Bradbury, R. Cooke, J. Erlandson, J.A. Estes, T.P. Hugues, S. Kidwell, C.B. Lange, H.S. Lenihan, J.M. Pandolfi, C.H. Peterson, R.S. Steneck, M.J. Tegner, and R.R. Warner. 2001. Historical overfishing and the recent collapse of coastal ecosystems. *Science* 293:629-637.

Jacobson, N., D. Butterill, and P. Goering. 2004. Organizational factors that influence university-based researchers' engagement in knowledge transfer activities. *Science Communication* 25(3):246-259.

Jaimes, A., N. Gaur, L.M.T. Aparecido, M.E. Everett, P. Knappett, M. Lawing, S. Majumder, G.R. Miller, G.W. Moore, C. Morgan, and B. Mitra. 2017. *Ecohydrological responses to Hurricane Harvey across South-Central Texas: A multidisciplinary approach of the Texas Water Observatory,* NH23E-2878. Presented at the 2017 American Geophysical Union Fall Meeting, New Orleans, LA.

Janasie, C.M., and S. Deal. 2015. *Increasing climate resilience on Dauphin Island through land use planning.* Mississippi-Alabama Sea Grant Program, MASGP-15-23, June 2015.

Jankowski, K.L., T.E. Törnqvist, and A.M. Fernandes. 2017. Vulnerability of Louisiana's coastal wetlands to present-day rates of relative sea-level rise. *Nature Communications* 8:14792. doi:10.1038/ncomms14792.

Jerome, S. 2017. *Hurricane season has left wastewater systems in disrepair water.* Water Online. September 28. Available at https://www.wateronline.com/doc/hurricane-season-has-left-wastewater-systems-in-disrepair-0001 [accessed May 16, 2018].

Jeter, Jr., G.W., and G.A. Carter. 2016. Habitat change on Horn Island, Mississippi, 1940–2010, determined from textural features in panchromatic vertical aerial imagery. *Geocarto International* 31(9):985-994. doi:10.1080/10106049.2015.1094527.

Johnson, D.R., J.R. Fischbach, and D.S. Ortiz. 2013. Estimating surge-based flood risk with the Coastal Louisiana Risk Assessment Model. *Journal of Coastal Research* 67(sp1):109-126.

Johnston, J.B., D.R. Cahoon, M.K. La Peyre. 2009. *Outer Continental Shelf (OCS)-related pipelines and navigation canals in the western and central Gulf of Mexico: Relative impacts on wetland habitats and effectiveness of mitigation.* U.S. Department of the Interior, Minerals Management Service, Gulf of Mexico OCS Region, New Orleans, LA. OCS Report MMS 2009-048.

Jones, C.E., and R.G. Blom. 2014. Bayou Corne, Louisiana, sinkhole: Precursory deformation measured by radar interferometry. *Geology* 42(2):111-114. doi:10.1130/G34972.1.

Jones, C.E., K. An, R.G. Blom, J.D. Kent, E.R. Ivins, and D. Bekaert. 2016. Anthropogenic and geologic influences on subsidence in the vicinity of New Orleans, Louisiana. *Journal of Geophysical Research: Solid Earth* 121(5):3867-3887. doi:10.1002/2015JB012636.

Kaniasty, K. 2012. Predicting social psychological well-being following trauma: The role of postdisaster social support. *Psychological Trauma: Theory, Research, Practice, and Policy* 4(1):22-33. doi: 10.1037/a0021412.

Karegar, M.A., T.H. Dixon, and R. Malservisi. 2015. A three-dimensional surface velocity field for the Mississippi Delta: Implications for coastal restoration and flood potential. *Geology* 43(6):519-522. doi:10.1130/g36598.1.

Kasmarek, M. 2013. *Hydrogeology and simulation of groundwater flow and land-surface subsidence in the northern part of the Gulf Coast aquifer system, Texas, 1891–2009.* U.S. Geological Survey Scientific Investigations Report 2012–5154.

Kellens, W., T. Terpstra, and P. De Maeyer, 2013. Perception and communication of flood risks: A systematic review of empirical research. *Risk Analysis* 33(1):24-49. doi:10.1111/j.1539-6924.2012.01844.x.

Kendall, M.S., T. Battista, and Z. Hillis-Starr. 2004. Long-term expansion of a deep *Syringodium filiforme* meadow in St. Croix, U.S. Virgin Islands: The potential role of hurricanes in the dispersal of seeds. *Aquatic Botany* 78:15-25.

Kennedy, A., S. Rogers, A. Sallenger, U. Gravois, B. Zachry, M. Dosa, and F. Zarama. 2011. Building destruction from waves and surge on the Bolivar Peninsula during Hurricane Ike. *Journal of Waterway, Port, Coastal, and Ocean Engineering* 137(3):132-141.

Kennedy, D. 2006. *Science Magazine's state of the planet, 2006–2007.* Washington, DC: Island Press.

Kennish, M.J. 1999. *Estuary restoration and maintenance: The national estuary program.* Boca Raton, FL: CRC Press.

Keown, M.P., E.A. Dardeau, Jr., and E.M. Causey. 1986. Historic trends in the sediment flow regime of the Mississippi River. *Water Resources Research* 22(11):1555-1564.

Khade, V., J. Kurian, P. Chang, I. Szunyogh, K. Thyng, and R. Montuoro. 2017. Oceanic ensemble forecasting in the Gulf of Mexico: An application to the case of the *Deepwater Horizon* oil spill. *Ocean Modelling* 113:171-184.

Khalil, S.M., A.M. Freeman, and R.C. Raynie. 2018. Sediment management for sustainable ecosystem restoration of coastal Louisiana. *Shore and Beach* 86(1):17-27.

Kim, C.-K., K. Park, S.P. Powers, W.M. Graham, and K.M. Bayha. 2010. Oyster larval transport in coastal Alabama: Dominance of physical transport over biological behavior in a shallow estuary. *Journal of Geophysical Research* 115:1-16.

Kirchhoff, C.J., M.C. Lemos, and S. Dessai. 2013. Actionable knowledge for environmental decision making: Broadening the usability of climate science. *Annual Review of Environment and Resources* 38:393-414. doi:10.1146/annurev-environ-022112-112828.

Kirwan, M.L., and J.P. Megonigal. 2013. Tidal wetland stability in the face of human impacts and sea-level rise. *Nature* 504(7478):53-60.

Kirwan, M.L., G.R. Guntenspergen, A. D'Alpaos, J.T. Morris, S.M. Mudd, and S. Temmerman. 2010. Limits on the adaptability of coastal marshes to rising sea level. *Geophysical Research Letters* 37(23). doi:10.1029/2010GL045489.

Kobayashi, N., A.W. Raichle, and T. Asano. 1993. Wave attenuation by vegetation. *Journal of Waterway, Port, Coastal, and Ocean Engineering* 119(1):30-48.

Koch, E.W., E.B. Barbier, B.R. Silliman, D.J. Reed, G.M. Perillo, S.D. Hacker, E.F. Granek, J.H. Primavera, N. Muthiga, and S. Polasky. 2009. Non-linearity in ecosystem services: Temporal and spatial variability in coastal protection. *Frontiers in Ecology and the Environment* 7(1):29-37.

Kolker, A.S., and S. Hameed. 2007. Meteorologically driven trends in sea level rise. *Geophysical Research Letters* 34(23).

Kolker, A.S., M.A. Allison, and S. Hameed. 2011. An evaluation of subsidence rates and sea-level variability in the northern Gulf of Mexico. *Geophysical Research Letters* 38(21). doi:10.1029/2011GL049458.

Kolker, A.S., M.D. Miner, and H.D. Weathers. 2012. Depositional dynamics in a river diversion receiving basin: The case of the West Bay Mississippi River diversion. *Estuarine, Coastal and Shelf Science* 106:1-12. doi:10.1016/j.ecss.2012.04.005.

Kolker, A.S., C. Li, N.D. Walker, C. Pilley, A.D. Ameen, G. Boxer, C. Ramatchandirane, M. Ullah, and K.A. Williams. 2014. The impacts of the Great Mississippi/Atchafalaya River flood on the oceanography of the Atchafalaya Shelf. *Continental Shelf Research* 86:17-33.

Kreibich, H., A.H. Thieken, T. Petrow, M. Müller, and B. Merz. 2005. Flood loss reduction of private households due to building precautionary measures—Lessons learned from the Elbe flood in August 2002. *Natural Hazards and Earth System Science* 5(1):117-126.

Kumpf, H., K. Steidinger, and K. Sherman (eds.). 1999. The Gulf of Mexico large marine ecosystem: Assessment, sustainability, and management. *Blackwell Science.*

Kunreuther, H. 1996. Mitigating disaster losses through insurance. *Journal of Risk and Uncertainty* 12(2-3):171-87. doi:10.1007/BF00055792.

Kwadijk, J.C.J., M. Haasnoot, J.P.M. Mulder, M.M.C. Hoogvliet, A.B.M. Jeuken, R.A.A. van der Krogt, N.G.C. van Oostrom, H.A. Schelfhout, E.H. van Velzen, H. van Waveren, and M.J.M. de Wit. 2010. Using adaptation tipping points to prepare for climate change and sea level rise: A case study in the Netherlands. *Wiley Interdisciplinary Reviews: Climate Change* 1(5):729-740. doi: 10.1002/wcc.64.

La Peyre, M.K., A.T. Humphries, S.M. Casas, and J.F. La Peyre. 2014. Temporal variation in development of ecosystem services from oyster reef restoration. *Ecological Engineering* 63:34-44.

Lackey, R.T. 2007. Science, scientists, and policy advocacy. *Conservation Biology* 21(1):12-17.

LACPRA (Louisiana Coastal Protection and Restoration Authority). 2012. *Louisiana's Comprehensive Master Plan for a Sustainable Coast.* Available at http://coastal.la.gov/a-common-vision/2012-coastal-master-plan [accessed May 21, 2018].

LACPRA. 2017. *Louisiana's Comprehensive Master Plan for a Sustainable Coast.* Available at http://coastal.la.gov/wp-content/uploads/2017/01/DRAFT-2017-Coastal-Master-Plan.pdf [accessed May 21, 2018].

Lambeck, K., H. Rouby, A. Purcell, Y. Sun, and M. Sambridge. 2014. Sea level and global ice volumes from the Last Glacial Maximum to the Holocene. *Proceedings of the National Academy of Sciences of the United States of America* 111(43):15296-15303. doi:10.1073/pnas.1411762111.

Landry, C.E., and M.R. Jahan-Parvar. 2011. Flood insurance coverage in the coastal zone. *Journal of Risk and Insurance* 78(2):361-388.

Larson, M. 1990. *SBEACH: Numerical model for simulating storm-induced beach change.* U.S. Army Corps of Engineers Technical Report CERC-89-9.

Leatherman, S.P. 1979. Barrier dune systems: A reassessment. *Sedimentary Geology* 24(1-2):1-16.

Lee, J.A., and R. Ignaciuk. 1985. The physiological ecology of strandline plants. In *Ecology of coastal vegetation* (pp. 319-326). Springer, Dordrecht.

Lee, T.M., E.M. Markowitz, P.D. Howe, C.Y. Ko, and A.A. Leiserowitz. 2015. Predictors of public climate change awareness and risk perception around the world. *Nature Climate Change* 5:1014-1020. doi:10.1038/nclimate2728.

Lehrter, J.C., D.L. Beddick, R. Devereux, D.F. Yates, and M.C. Murrell. 2012. Sediment-water fluxes of dissolved inorganic carbon, O2, nutrients, and N2 from the hypoxic region of the Louisiana continental shelf. *Biogeochemistry* 109(1):233-252.

Lellis-Dibble, K.A., K.E. McGlynn, and T.E. Bigford. 2008. *Estuarine fish and shellfish species in U.S. commercial and recreational fisheries: Economic value as an incentive to protect and restore estuarine habitat.* Silver Spring, MD: U.S. Department of Commerce, NOAA Technical Memo. NMFS-F/SPO-90.

Lemasson, A., S. Fletcher, J.M. Hall-Spencer, and A.M. Knights. 2017. Linking the biological impacts of ocean acidification on oysters to changes in ecosystems services: A review. *Journal of Experimental Marine Biology and Ecology* 492:49-62.

Lempert, R.J., S.W. Popper, and S.C. Bankes. 2003. *Shaping the next one hundred years: New methods for quantitative, long-term policy analysis.* Santa Monica, CA: RAND Corporation, MR-1626-RPC.

Lesser, G., J.V. Roelvink, J. Van Kester, and G. Stelling. 2004. Development and validation of a three-dimensional morphological model. *Coastal Engineering* 51(8-9):883-915.

Letetrel, C., M. Karpytchev, M.-N. Bouin, M. Marcos, A. Santamaría-Gómez, and G. Wöppelmann. 2015. Estimation of vertical land movement rates along the coasts of the Gulf of Mexico over the past decades. *Continental Shelf Research* 111:42-51. doi:10.1016/j.csr.2015.10.018.

Lin, N., K. Emanuel, M. Oppenheimer, and E. Vanmarcke. 2012. Physically based assessment of hurricane surge threat under climate change. *Nature Climate Change* 2(6):462-467.

Lipiec, E., P. Ruggiero, A. Mills, K. Serafin, J. Bolte, P. Corcoran, J. Stevenson, and C. Zanocco. 2018. Mapping out climate change: Assessing how coastal communities adapt using alternative future scenarios. *Journal of Coastal Research* (In press). doi: 10.2112/JCOASTRES-D-17-00115.1.1.

Liu, J., T. Dietz, S.R. Carpenter, M. Alberti, C. Folke, E. Moran, A.N. Pell, P. Deadman, T. Kratz, J. Lubchenco, and E. Ostrom. 2007. Complexity of coupled human and natural systems. *Science* 317(5844):1513-1516.

Liu, Y., R.H. Weisberg, S. Vignudelli, and G.T. Mitchum. 2016. Patterns of the Loop Current system and regions of sea surface height variability in the eastern Gulf of Mexico revealed by the self-organizing maps. *Journal of Geophysical Research: Oceans* 121(4):2347-2366.

Lopez, A., B. Roberts, D. Heimiller, N. Blair, and G. Porro. 2012. *U.S. renewable energy technical potentials: A GIS-based approach.* National Renewable Energy Laboratory Technical Report, NREL/TP-6A20-51946.

Lopez, J.A. 2009. The multiple lines of defense strategy to sustain coastal Louisiana. *Journal of Coastal Research: Special Issue* 54:186-197.

Lorenzo-Trueba, J., and A.D. Ashton. 2014. Rollover, drowning, and discontinuous retreat: Distinct modes of barrier response to sea-level rise arising from a simple morphodynamic model. *Journal of Geophysical Research: Earth Surface* 119(4):779-801.

Love, R., G.A. Milne, L. Tarasov, S.E. Engelhart, M.P. Hijma, K. Latychev, B.P. Horton, and T.E. Törnqvist. 2016. The contribution of glacial isostatic adjustment to projections of sea level change along the Atlantic and Gulf coasts of North America. *Earth's Future* 4(10):440-464.

Lucas, K.L., and G.A. Carter. 2010. Decadal-scale changes in habitat abundance and distribution on Horn Island, Mississippi. *Journal of Coastal Research* 26:1142-1148. doi:10.2112/JCOASTRES-D-09-00018.1.

Lucas, K.L., and G.A. Carter. 2013. Change in distribution and composition of vegetated habitats on Horn Island, Mississippi, northern Gulf of Mexico, in the initial five years following Hurricane Katrina. *Geomorphology* 199:129-137. doi:10.1016/j.geomorph.2012.11.010.

Lugo-Fernández, A., H.H. Roberts, and J.N. Suhayda. 1998. Wave transformations across a Caribbean fringing-barrier coral reef. *Continental Shelf Research* 18(10):1099-1124.

Macreadie, P.I., M.E. Baird, S.M. Trevathan-Tackett, A.W.D. Larkum, and P.J. Ralph. 2014. Quantifying and modelling the carbon sequestration capacity of seagrass meadows – A critical assessment. *Marine Pollution Bulletin* 83(2):430-439.

Mallinson, D.J., C.W. Smith, S.J. Culver, S.R. Riggs, and D. Ames. 2010. Geological characteristics and spatial distribution of paleo-inlet channels beneath the outer banks barrier islands, North Carolina, USA. *Estuarine, Coastal and Shelf Science* 88(2):175-189.

Manis, J.E., S.K. Garvis, S.M Jachec and L.J. Walters. 2015. Wave attenuation experiments over living shorelines over time: A wave tank study to assess recreational boating pressures. *Journal of Coastal Conservation* 19(1):1-11.

Marcy, D.C., A.L. Allen, W. Sweet, S.K. Gill, A. Luscher-Aissaoui, E.P. Myers, and C. Eugene Zervas. 2012. *Incorporating sea level change scenarios at the local level.* Available at https://www.ngs.noaa.gov/PUBS_LIB/SLCScenariosLL.pdf [accessed May 24, 2018].

Martin, S.K., G. Savant, and D.C. McVan. 2010. *Lake Borgne surge barrier study.* U.S. Army Corps of Engineers, Engineer Research and Development Center, Technical Report; ERDC/CHL TR-10-10.

Martínez, M.L., A.M. Maun, and N.P. Psuty. 2004. The fragility and conservation of the world's coastal dunes: Geomorphological, ecological and socioeconomic perspectives (pp. 355-370). In M.L. Martínez and N. Psuty (eds.). *Coastal dunes: Ecology and conservation.* Berlin: Springer-Verlag.

Masetti, R., S. Fagherazzi, and A. Montaranri. 2008. Application of a barrier island translation model to the millenial-scale evolution of Sand Key, Florida. *Continental Shelf Research* 28(9):1116-1126.

Maun, M.A. 1998. Adaptations of plants to burial in coastal sand dunes. *Canadian Journal of Botany* 76(5):713-738. doi:10.1139/b98-058.

McDaniels, T.L., R.S. Gregory, and D. Fields. 1999. Democratizing risk management: Successful public involvement in local water management decisions. *Risk Analysis* 19(3):497-510. doi: 10.1023/A:1007060931193.

McDonald, A., P. Prado, K. Heck, J.W. Fourqurean, T.A. Frankovich, K.H. Dunton, and J. Cebrian. 2016a. Seagrass growth, reproductive and morphological plasticity across environmental gradients over a large spatial scale. *Aquatic Botany* 134:87-96.

McDonald, R.B., R.M. Moody, K.L. Heck, and J. Cebrian. 2016b. Fish, macroinvertebrate and epifaunal communities in shallow coastal lagoons with varying seagrass cover of the northern Gulf of Mexico. *Estuaries and Coasts* 39(3):718-730.

McLeman, R. 2017. Thresholds in climate migration. *Population and Environment* 1-20. doi:10.1007/s11111-017-0290-2.

McLeman, R., and B. Smit. 2006. Migration as an adaptation to climate change. *Climatic Change* 76(1-2):31-53.

McNamara, D.E., and A. Keeler. 2013. A coupled physical and economic model of the response of coastal real estate to climate risk. *Nature Climate Change* 3(6):559-562.

McNamara, D.E., and B.T. Werner. 2008. Coupled barrier island–resort model: 1. Emergent instabilities induced by strong human-landscape interactions. *Journal of Geophysical Research: Earth Surface* 113(F1).

McNamara, D.E., S. Gopalakrishnan, M.D. Smith, and A.B. Murray. 2015. Climate adaptation and policy-induced inflation of coastal property value. *PLOS ONE* 10(3):p.e0121278.

Meade, R.H., and J.A. Moody. 2010. Causes for the decline of suspended-sediment discharge in the Mississippi River system, 1940–2007. *Hydrological Processes* 24(1):35-49.

Mendez, F.J., and I.J. Losada. 2004. An empirical model to estimate the propagation of random breaking and nonbreaking waves over vegetation fields. *Coastal Engineering* 51(2):103-118. doi:10.1016/j.coastaleng.2003.11.003.

Meselhe, E., J.A. McCorquodale, J. Shelden, M. Dortch, T.S. Brown, P. Elkan, M.D. Rodrigue, J.K. Schindler, and Z. Wang. 2013. Ecohydrology component of Louisiana's 2012 Coastal Master Plan: Mass-balance compartment model. *Journal of Coastal Research* 67(sp1):16-28.

Meyer, M.A., B. Mitchell, J.C. Purdum, K. Breen, and R.L. Iles. 2018. Previous hurricane evacuation decisions and future evacuation intentions among residents of Southeast Louisiana. *International Journal of Disaster Risk Reduction* (In press).

Michel-Kerjan, E.O., and C. Kousky. 2010. Come rain or shine: Evidence on flood insurance purchases in Florida. *Journal of Risk and Insurance* 77(2):369-397.

Milliken, K.T., J.B. Anderson, A.R. Simms, and M.D. Blum. 2017. A Holocene record of flux of alluvial sediment related to climate: Case studies from the northern Gulf of Mexico. *Journal of Sedimentary Research* 87(8):780-794.

Miner, M.D., M.A. Kulp, D.M. FitzGerald, and I.Y. Georgiou. 2009. Hurricane-associated ebb-tidal delta sediment dynamics. *Geology* 37(9):851-854.

Mitrovica, J.X., N. Gomez, and P.U. Clark. 2009. The sea-level fingerprint of West Antarctic collapse. *Science* 323(5915):753.

Moody, R.M., J. Cebrian, and K.L. Heck, Jr. 2013. Interannual recruitment dynamics for resident and transient marsh species: Evidence for a lack of impact by the Macondo oil spill. *PLOS ONE* 8(3):p.e58376.

Moore, L.J., J.H. List, S.J. Williams, and D. Stolper. 2010. Complexities in barrier island response to sea-level rise: Insights from model experiments. *Journal of Geophysical Research: Earth Surface* 115(F3). doi:10.1029/2009JF001299.

Moore, L.J., D.E. McNamara, A.B. Murray, and O. Brenner. 2013. Observed changes in hurricane-driven waves explain the dynamics of modern cuspate shorelines. *Geophysical Research Letters* 40(22):5867-5871.

Moore, L.J., K. Patsch, S.J. Williams, and J.L. List. 2014. Barrier islands poised for geomorphic threshold crossing in response to rapid sea-level rise: Insights from numerical model experiments, Chandeleur Islands, Louisiana, USA. *Marine Geology* 355(1):244-259. doi:10.1016/j.margeo.2014.05.022.

Morang, A. 2006. North Texas sediment budget. U.S. Army Corps of Engineers, Engineer Research and Development Center, Technical Report; ERDC/CHL TR-06-17.

Morris, J.T., P.V. Sundareshwar, C.T. Nietch, B. Kferfve, and D.R. Cahoon. 2002. Responses of coastal wetlands to rising sea level. *Ecology* 83:2869-2877. doi:10.1890/0012-9658(2002)083[2869:ROCWTR]2.0.CO;2.

Morton, R.A., 2008. Historical changes in the Mississippi-Alabama barrier island chain and the roles of extreme storms, sea level, and human activities. *Journal of Coastal Research* 24(6):1587-1600.

Morton, R.A., and J.C. Bernier. 2010. Recent subsidence-rate reductions in the Mississippi Delta and their geological implications. *Journal of Coastal Research* 26(3):555-561. doi:10.2112/JCOASTRES-D-09-00014R1.1.

Morton, R.A., T.L. Miller, and L.J. Moore. 2004. *National assessment of shoreline change: Part 1 historical shoreline changes and associated coastal land loss along the U.S. Gulf of Mexico*. U.S. Geological Survey. OFR 2004-1043. Available at http://pubs.usgs.gov/of/2004/1043 [accessed May 21, 2018].

Morton, R.A., T.L. Miller, and L.J. Moore. 2005. Historical shoreline changes along the U.S. Gulf of Mexico: A summary of recent shoreline comparisons and analyses. *Journal of Coastal Research* 21(4):704-709.

Morton, R.A., J.C. Bernier, and K.W. Kelso. 2009. *Recent subsidence and erosion at diverse wetland sites in the southeastern Mississippi Delta plain* (No. 2009-1158). U.S. Geological Survey.

Murray, G.E. 1961. *Geology of the Atlantic and Gulf coastal provinces of North America*. New York: Harper.

Myers, C.A., T. Slack, and J. Singelmann. 2008. Social vulnerability and migration in the wake of disaster: The case of Hurricanes Katrina and Rita. *Population and Environment* 29(6):271-291.

Nairn, R.B., Q. Lu, and S.K. Langendyk. 2005. *A study to address the issue of seafloor stability and the impact on oil and gas infrastructure in the Gulf of Mexico*. U.S. Department of the Interior, Minerals Management Service, Gulf of Mexico OCS Region, New Orleans, LA. OCS Study MMS 2005-043.

Narayan, S., M.W. Beck, B.G. Reguero, I.J. Losada, B. Van Wesenbeeck, N. Pontee, J.N. Sanchirico, J.C. Ingram, G.M. Lange, and K.A. Burks-Copes. 2016. The effectiveness, costs and coastal protection benefits of natural and nature-based defences. *PLOS ONE* 11(5):e0154735.

NASA (National Aeronautics and Space Adminstration). 2016. *Surface Water and Ocean Topography Mission (SWOT) project: Science requirements*. Jet Propulsion Laboratory. Available at https://swot.jpl.nasa.gov/docs/D-61923_SRD_Rev_A_20160318.pdf [accessed May 21, 2018].

NASEM (National Academies of Sciences, Engineering, and Medicine). 2017a. *Communicating science effectively: A research agenda*. Washington, DC: The National Academies Press.

NASEM. 2017b. *Effective monitoring to evaluate ecological restoration in the Gulf of Mexico*. Washington, DC: The National Academies Press.

NASEM. 2018. *Understanding and predicting the Gulf of Mexico Loop Current: Critical gaps and recommendations*. Washington, DC: The National Academies Press.

National Hurricane Center, National Oceanic and Atmospheric Administration. 2018. Costliest U.S. tropical cyclones tables updated. Available at https://www.nhc.noaa.gov/news/UpdatedCostliest.pdf [accessed May 21, 2018].

National Ocean Service, National Oceanic and Atmospheric Administration. 2011. *NOAA's state of the coast. The Gulf of Mexico at a Glance: A Second Glance*. Washington, DC: U.S. Department of Commerce.

Needham, H., and L. Carter. 2012. *Gulf Coast climate information needs assessment*. Southern Climate Impacts Planning Program (SCIPP). Available at http://www.southernclimate.org/publications/Gulf_Coast_Assessment_Final.pdf [accessed May 21, 2018].

Needham, H.F., and B.D. Keim. 2012. A storm surge database for the U.S. Gulf Coast. *International Journal of Climatology* 32(14):2108-2123.

Needham, H.F., D.P. Brown, and L.M. Carter. 2012. *Impacts and adaptation options in the Gulf Coast*. Center for Climate and Energy Solutions. Available at https://www.c2es.org/site/assets/uploads/2012/05/gulf-coast-impacts-adaptation.pdf [accessed May 21, 2018].

Nerem, R.S., B.D. Beckley, J.T. Fasullo, B.D. Hamlington, D. Masters, and G.T. Mitchum. 2018. Climate change–driven accelerated sea-level rise detected in the altimeter era. *Proceedings of the National Academy of Sciences of the United States of America* 115(9):2022-2025. doi:10.1073/pnas.1717312115.

Newstrom, J. W., and K. Davis. 1986. *Human behavior at work*. New York, NY: McGraw-Hill Education.

Nienhuis, J.H., T.E. Törnqvist, K.L. Jankowski, A.M. Fernandes, and M.E. Keogh. 2017. A new subsidence map for coastal Louisiana. *GSA Today* 27:58-59.

NIST (National Institute of Standards and Technology). 2006. *Performance of physical structures in hurricane Rita and hurricane Katrina: A reconnaissance report*. National Institute of Standards and Technology (NIST) Technical Note 1476.

NOAA (National Oceanic and Atmospheric Administration). 2013. *National coastal population report: Population trends from 1970 to 2020*. NOAA State of the Coast Report Series. Available at https://coast.noaa.gov/digitalcoast/training/population-report.html [accessed May 21, 2018].

NOAA. 2017. *2017 ecosystem status report update for the Gulf of Mexico*. NOAA Technical Memorandum NMFS-SEFSC-706. National Marine Fisheries Service. Southeast Fisheries Science Center. Miami, FL. Available at http://www.aoml.noaa.gov/ocd/ocdweb/ESR_GOMIEA/report/GoM_EcosystemStatusReport2017.pdf [accessed May 21, 2018].

NOAA. 2018. Tides and Currents. Available at https://tidesandcurrents.noaa.gov [accessed May 21, 2018].

NOS (National Ocean Service), National Oceanic and Atmospheric Administration. 2011. *NOAA's State of the Coast. The Gulf of Mexico at a glance: A second glance*. Washington, DC: U.S. Department of Commerce.

NRC (National Research Council). 1994. *Improving the safety of marine pipelines*. Washington, DC: National Academy Press.

NRC. 2009. *Informing decisions in a changing climate*. Washington, DC: The National Academies Press.

NRC. 2013. *Water resources infrastructure: Deterioration, investment, or divestment?* Washington, DC: The National Academies Press.

NREL (National Renewable Energy Laboratory). 2016. *2016 standard scenarios report: A U.S. electricity sector outlook.* Technical Report NREL/TP-6A20-66939.

Oliver-Smith, A. 1996. Anthropological research on hazards and disasters. *Annual Review of Anthropology* 25(1):303-328.

Oosting, H.J. 1954. Ecological processes and vegetation of the maritime strand in the southeastern United States. Botanical Review 20:226-262.

Osland, M.J., N. Enwright, R.H. Day, and T.W. Doyle. 2013. Winter climate change and coastal wetland foundation species: Salt marshes vs. mangrove forests in the southeastern United States. *Global Change Biology* 19:1482-1494.

Osland, M.J., K.T. Griffith, J.C. Larriviere, L.C. Feher, D.R. Cahoon, N.M. Enwright, D.A. Oster, J.M. Tirpak, M.S. Woodrey, R.C. Collini, and J.J. Baustian. 2017. Assessing coastal wetland vulnerability to sea-level rise along the northern Gulf of Mexico coast: Gaps and opportunities for developing a coordinated regional sampling network. *PLOS ONE* 12(9). doi:p.e0183431.

Otvos, E.G. 2012. Coastal barriers—Nomenclature, processes, and classification issues. *Geomorphology* 39:39-52.

Otvos, E.G., and G.A. Carter. 2008. Hurricane degradation—Barrier development cycles, northeastern Gulf of Mexico: Landform evolution and island chain history. *Journal of Coastal Research* 463-478.

Otvos, E.G., and G.A. Carter. 2013. Regressive and transgressive barrier islands on the north-central Gulf Coast: Contrasts in evolution, sediment delivery, and island vulnerability. *Geomorphology* 198:1-19. doi:10.1016/j.geomorph.2013.05.015.

Pabody, C.M., R.H. Carmichael, L. Rice, and M. Ross. 2009. A new sighting network adds to 20 years of historical data on fringe West Indian manatee (*Trichechus manatus*) populations in Alabama waters. *Gulf of Mexico Science* 27(1):52-61.

Padgett, J., R. DesRoches, B. Nielson, M. Yashinsky, O.S. Kwon, N. Burdette, and E. Tavera. 2008. Bridge damage and repair costs from Hurricane Katrina. *Journal of Bridge Engineering* 13(1):6-14.

Paine, J.G., 1993. Subsidence of the Texas coast: Inferences from historical and late Pleistocene sea levels. *Tectonophysics* 222:445-458.

Palmsten, M.L., K.L. Brodie, and N.J. Spore. 2017. *Coastal foredune evolution, part 2: Modeling approaches for meso scale morphologic evolution.* Naval Research Laboratory Stennis Space Center.

Panchang, A.J., C.K. Jeong, and Z. Demirbilek. 2013. Analysis of extreme wave heights in the Gulf of Mexico for offshore engineering applications. *Journal of Offshore Mechanics and Artic Engineering* 135(3):031104.

Paola, C. 2001. *Modelling stream braiding over a range of scales.* New Zealand Hydrological Society. Available at http://hdl.handle.net/11299/164368 [accessed May 21, 2018].

Paola, C., R.R. Twilley, D.A. Edmonds, W. Kim, D. Mohrig, G. Parker, E. Viparelli, and V.R. Voller. 2011. Natural processes in delta restoration: Application to the Mississippi Delta. *Annual Review of Marine Science* 3:67-91.

Pape, L., Y. Kuriyama and B.G. Ruessink. 2010. Models and scales for cross-shore sand bar migration. *Journal of Geophysical Research* 115(F3). doi:10.1029/2009JF001644.

Passeri, D.L., J.W. Long, N..G. Plant, M.V. Bilskie, and S.C. Hagen. 2018. The influence of bed friction variability due to land cover on storm-driven barrier island morphodynamics. *Coastal Engineering* 132:82-94.

Pendleton, E.A., J.A. Barras, S.J. Williams, and D.C. Twichell. 2010. *Coastal vulnerability assessment of the Northern Gulf of Mexico to sea-level rise and coastal change: U.S. Geological Survey Open-File Report 2010-1146.* Available at https://pubs.usgs.gov/of/2010/1146 [accessed May 21, 2018].

Penland, S., and K.E. Ramsey. 1990. Relative sea-level rise in Louisiana and the Gulf of Mexico: 1908–1988. *Journal of Coastal Research* 6(2):323-342.

Penland, S., J.R. Suter, R. Boyd. 1985. Barrier island arcs along abandoned Mississippi River deltas. *Marine Geology* 63(1-4):197-233.

Penland, S., R. Boyd, and J.R. Suter. 1988. Transgressive depositional systems of the Mississippi Delta plain: A model for barrier shoreline and shelf sand development. *Journal of Sedimentary Petrology* 58(6):932-949.

Penland, S., L. Wayne, L.D. Britsch, S.J. Williams, and V.C. Butterworth. 2000. *Process classification of coastal land loss between 1932 and 1990 in the Mississippi River Delta Plain, Southeastern Louisiana.* U.S. Geological Survey Open File Report, 00-418.

Perrin, A., and M. Duggan. 2015. *Americans' Internet access: 2000–2015.* Washington, DC: Pew Research Center.

Peyronnin, N., M. Green, C.P. Richards, A. Owens, D. Reed, J. Chamberlain, D.G. Groves, W.K. Rhinehart, and K. Belhadjali. 2013. Louisiana's 2012 Coastal Master Plan: Overview of a science-based and publicly informed decision-making process. *Journal of Coastal Research* 67(sp1):1-5.

Pickering, M.D., K.J. Horsburgh, J.R. Blundell, J.J.-M. Hirschi, R.J. Nichols, M. Verlan, and N.C. Wells. 2017. The impact of future sea-level rise on the global tides. *Continental Shelf Research* 142:50-68.

Picou, J.S., and B.K. Marshall. 2007. Katrina as paradigm shift: Reflections on disaster research in the 21st century. In *The sociology of Katrina: Perspectives on a modern catastrophe* (pp. 1-20).

Pilkey, O.H., 2003. *A celebration of the world's barrier islands*. New York: Columbia University Press.

Pistrika, A.K., and S.N. Jonkman. 2010. Damage to residential buildings due to flooding of New Orleans after hurricane Katrina. *Natural Hazards* 54(2):413-434.

Priest, T. 2007. Extraction not creation: The history of offshore petroleum in the Gulf of Mexico. *Enterprise & Society* 8(2):227-267.

Prokopy, L.S., J.S. Carlton, T. Haigh, M.C. Lemos, A.S. Mase, and M. Widhalm. 2017. Useful to usable: Developing usable climate science for agriculture. *Climate Risk Management* 15:1-7.

PSDS (Program for the Study of Developed Shorelines at Western Carolina University). 2018. Beach Nourishment Viewer. Available at http://beachnourishment.wcu.edu [accessed May 3, 2018].

Psuty, N.P. 2008. The coastal foredune: A morphological basis for regional coastal dune development. In M.L. Martínez and N.P. Psuty (eds), *Coastal Dunes. Ecological Studies*, vol. 171. Berlin, Heidelberg: Springer.

Pugh, D.T., 1987. *Tides, surges and mean sea-level: A handbook for engineers and scientists*. Chichester: John Wiley & Sons.

Reguero, B.G., M.W. Beck, D.N. Bresch, J. Calil, and I. Meliane. 2018. Comparing the cost effectiveness of nature-based and coastal adaptation: A case study from the Gulf Coast of the United States. *PLOS ONE* 13(4). doi:p.e0192132.

Renn, O., and D. Levine. 1991. Credibility and trust in risk communication. In *Communicating risks to the public* (pp. 175-217). Dordrecht: Springer.

Riad, J.K., F.H. Norris, and R.B. Ruback. 1999. Predicting evacuation in two major disasters: Risk perception, social influence, and access to resources. *Journal of Applied Social Psychology* 29(5):918-934.

Rietbroek, R., S.E. Brunnabend, J. Kusche, J. Schröter, and C. Dahle. 2016. Revisiting the contemporary sea-level budget on global and regional scales. *Proceedings of the National Academy of Sciences of the United States of America* 113(6):1504-1509. doi:10.1073/pnas.1519132113.

Roberts, H.H., O.K. Huh, S.A. Hsu, L.J. Rouse, and D.A. Rickman. 1989. Winter storm impacts on the chenier plain coast of southwestern Louisiana. *Gulf Coast Association of Geological Societies Transactions* 39:515-522.

Roberts, H.H., R.D. DeLaune, J.R. White, C. Li, C.E. Sasser, D. Braud, E. Weeks, and S. Khalil. 2015. Floods and cold front passages: Impacts on coastal marshes in a river diversion setting (Wax Lake Delta Area, Louisiana). *Journal of Coastal Research* 31(5):1057-1068.

Rodriguez, A.B., J.B. Anderson, F.P. Siringan, and M. Taviani. 2004. Holocene evolution of the east Texas coast and inner continental shelf: Along-strike variability in coastal retreat rates. *Journal of Sedimentary Research* 74(3):405-421.

Roelvink, D., A. Reniers, A.P. Van Dongeren, J.V.T. de Vries, R. McCall, and J. Lescinski. 2009. Modeling storm impacts on beaches, dunes, and barrier islands. *Coastal Engineering* 56(11):1133-1152.

Roelvink, J.A. 2006. Coastal morphodynamic evolution techniques. *Coastal Engineering* 53(2):277-287.

Rogers, L., L.J. Moore, E.B. Goldstein, C. Hein, J. Lorenzo-Trueba, and A. Ashton. 2015. Anthropogenic controls on overwash deposition: Evidence and consequences. *Journal of Geophysical Research: Earth Surface* 120(12):2609-2624.

Roland, R.M., and S.L. Douglas. 2005. Estimating wave tolerance of *Spartina alterniflora* in coastal Alabama. *Journal of Coastal Research* 21(3):453-463.

Roman, C.T., and K.F. Nordstrom. 1988. The effect of erosion rate on vegetation patterns of an East Coast barrier island. *Estuarine, Coastal and Shelf Science* 26(3):233-242.

Ruessink, B.G., Y. Kuriyama, A.J.H.M. Reniers, J.A. Roelvink, and D.J.R. Walstra. 2007. Modeling cross-shore sandbar behavior on the timescale of weeks. *Journal of Geophysical Research* 112(F3). doi:10.1029/2006JF000730.

Sala, E., O. Aburto-Oropeza, G. Paredes, I. Parra, J.C. Barrera, and P.K. Dayton. 2002. A general model for designing networks of marine reserves. *Science* 298:1991-1993.

Sallenger, Jr., A.H. 2000. Storm impact scale for barrier islands. *Journal of Coastal Research* 16(3):890-895.

Samuels, P., F. Klijn, and J. Dijkman. 2006. An analysis of the current practice of policies on river flood risk management in different countries. *Irrigation and Drainage* 55(S1):S141-S150. doi:10.1002/ird.257.

Sandmann, L.R., J.W. Jordan, C.D. Mull, and T. Valentine. 2014. Measuring boundary-spanning behaviors in community engagement. *Journal of Higher Education Outreach and Engagement* 18(3):83-96.

Santos, I.R., B.D. Eyre, and M. Huettel. 2012. The driving forces of porewater and groundwater flow in permeable coastal sediments: A review. *Estuarine, Coastal and Shelf Science* 98:1-15. doi:10.1016/j.ecss.2011.10.024.

Schanze, J., G. Hutter, E. Penning-Rowsell, H.P. Nachtnebel, V. Meyer, A. Werritty, T. Harries, H. Holzmann, B. Jessel, P. Koeniger, and C. Kuhlicke. 2008. *Systematisation, evaluation and context conditions of structural and non-structural measures for flood risk reduction.* FLOOD-ERA Joint Report. ERA-NET CRUE. London.

Scheffel, W.A., K.L. Heck, J. Cebrian, M. Johnson, and D. Byron. 2013. Range expansion of black mangroves (*Avicennia germinans*) to the Mississippi barrier islands. *Gulf of Mexico Science* 31:79-82.

Scheffer, M., J. Bascompte, W.A. Brock, V. Brovkin, S.R. Carpenter, V. Dakos, H. Held, E.H. van Nes, M. Rietkerk, and G. Sugihara. 2009. Early-warning signals for critical transitions. *Nature* 461:53.

Schneider, J.A., B. Cornwell, D. Ostrow, S. Michaels, P. Schumm, E.O. Laumann, and S. Friedman. 2013. Network mixing and network influences most linked to HIV infection and risk behavior in the HIV epidemic among black men who have sex with men. *American Journal of Public Health* 103(1):e28-e36.

Scyphers, S.B., S.P. Powers, J.L. Akins, J.M. Drymon, C.W. Martin, Z.H. Schobernd, P.J. Schofield, R.L. Shipp, and T.S. Switzer. 2015. The role of citizens in detecting and responding to a rapid marine invasion. *Conservation Letters* 8(4):242-250.

Sea Grant. 2018. Sea Grant in the Gulf of Mexico. Available at https://gulfseagrant.org [accessed March 20, 2018].

Shao, G., H.H. Shugart, and B.P. Hayden. 1996. Functional classifications of coastal barrier island vegetation. *Journal of Vegetation Science* 7(3):391-396.

Sharma, S., J. Goff, J. Cebrian and C. Ferraro. 2016a. A hybrid shoreline stabilization technique: Impact of modified intertidal reefs on marsh expansion and nekton habitat in the northern Gulf of Mexico. *Ecological Engineering* 90:352-360.

Sharma, S., J. Goff, R. Moody, D. Byron, K.L. Heck, S.P. Powers, C. Ferraro and J. Cebrian. 2016b. Do restored oyster reefs benefit seagrasses? An experimental study in the northern Gulf of Mexico. *Restoration Ecology* 24:306-313.

Sharma, S., J. Goff, R.M. Moody, A. McDonald, D. Byron, K.L. Heck Jr, S.P. Powers, C. Ferraro, and J. Cebrian. 2016c. Effects of shoreline dynamics on saltmarsh vegetation. *PLOS ONE* 11(7). doi: p.e0159814.

Shen, Z., N.H. Dawers, T.E. Törnqvist, N.M. Gasparini, M.P. Hijma, and B. Mauz. 2017. Mechanisms of late Quaternary fault throw-rate variability along the north-central Gulf of Mexico coast: Implications for coastal subsidence. *Basin Research* 29(5):557-570. doi:10.1111/bre.12184.

Shennan, I., and P.L. Woodworth. 1992. A comparison of late Holocene and 20th century sea-level trends from the U.K. and North Sea region. *Geophysical Journal International* 109(1):96-105.

Shepard, C.C., C.M. Crain, and M.W. Beck. 2011. The protective role of coastal marshes: A systematic review and meta-analysis. *PLOS ONE* 6(11):e27374. doi:10.1371/journal.pone.0027374.

Sherman, D.J., B.U. Hales, M.K. Potts, J.T. Ellis, H. Liu, and C. Houser. 2013. Impacts of Hurricane Ike on the beaches of the Bolivar Peninsula, TX, USA. *Geomorphology* 199:62-81.

Sherrieb, K., F.H. Norris, and S. Galea. 2010. Measuring capacities for community resilience. *Social Indicators Research* 99:227-247.

Short, A.D., and P.A. Hesp. 1982. Wave, beach, and dune interactions in southeastern Australia. *Marine Geology* 48:259-284.

Silvertown, J. 2009. A new dawn for citizen science. *Trends in Ecology and Evolution* 24(9):467-471.

Simeonov, J.A. and J. Calantoni. 2011. A pressure boundary integral method for direct fluid–particle simulations on Cartesian grids. *Journal of Computational Physics* 230(5):1749-1765.

Slott, J., A.B. Murray, A. Ashton, and T. Crowley. 2006. Coastline responses to changing storm patterns, *Geophysical Research Letters: Special Collection* 33(1). doi:10.1029/2006GL027445.

Slott, J.M., M.D. Smith, and A.B. Murray. 2008. Synergies between adjacent beach-nourishing communities in a morpho-economic coupled coastline model. *Coastal Management* 36:374-391.

Small, M.J., and S. Xian. 2018. Extreme event outcomes, scientific study, and implications for decision maker risk perception and mitigation. *Global Environmental Change* (under revision).

Smith, J., S. Banik, and U. Haque. 2018. Catastrophic hurricanes and public health dangers: Lesson learned. *Journal of Public Health and Emergency* 2(2).

Smith, M.D. 2008. Bioeconometrics: Empirical modeling of bioeconomic systems. *Marine Resource Economics* 23(1):1-23.

Smith, M.D. 2014. Fauna in decline: Management risks. *Science* 346(6211):819.

Smith, M.D., J.M Slott., D. McNamara, and A.B. Murray. 2009. Beach nourishment as a dynamic capital accumulation problem. *Journal of Environmental Economics and Management* 58(1):58-71.

Smith, M.D., J. Lynham, J.N. Sanchirico, and J. Wilson. 2010. The political economy of marine reserves: Understanding the role of opportunity costs. *Proceedings of the National Academy of Sciences of the United States of America* 107(43):18300-18305.

Smith, M.D., A.B. Murray, S. Gopalakrishnan, A.G. Keeler, C.E. Landry, D. McNamara, L. Moore. 2015. Geoengineering coastlines? From accidental to intentional (pp. 99-122). In J. Baztan, O. Chouinard, B. Jorgensen, P. Tett, J-P Vanderlinden, and L. Vasseuer (eds.). *Coastal zones: Solutions for the 21st century*. Amsterdam: Elsevier.

Snowden, J. O., W.C. Ward, and J.R.J. Studlick. 1980. Geology of greater New Orleans: Its relationship to land subsidence and flooding. *New Orleans Geological Society*.

Somerville, R.C., and S.J. Hassol. October 2011. Communicating the science of climate change. *Physics Today* 48-53.

Songy, G. 2017. *Beach nourishment along Galveston seawall: Completion of Texas' largest-ever beach nourishment*. ASBPA 2017 Symposium. Port Aransas, TX.

Sparks, E.L., J. Cebrian, P.D. Biber, K.L. Sheehan, and C.R. Tobias. 2013. Cost-effectiveness of two small-scale salt marsh restoration designs. *Ecological Engineering* 53:250-256.

Sparks, E.L., J. Cebrian, C.R. Tobias, and C. May. 2015. Groundwater nitrogen processing in northern Gulf of Mexico restored marshes. *Journal of Environmental Management* 150:206-215.

Speyrer, N., and A. Gaharan. 2017. 2017 *Coastal Master Plan: Appendix G: Outreach and engagement. Version final*. (pp. 1-31). Baton Rouge, LA: Coastal Protection and Restoration Authority.

Stallins, J.A., and A.J. Parker. 2003. The influence of complex systems interactions on barrier island dune vegetation pattern and process. *Annals of the Association of American Geographers* 93(1):13-29.

Stephens, J.C., L.H. Allen, and E. Chen. 1984. Organic soil subsidence. In Holzer, T.L., *Man-induced land subsidence: GSA Reviews in Engineering Geology* 6:107-122. doi:10.1130/REG6-p107.

Stern, R.J., and W.R. Dickinson. 2010. The Gulf of Mexico is a Jurassic backarc basin. *Geosphere* 6(6):739-754. doi:10.1130/GES00585.

Stewart, S. 2004. *Tropical cyclone report: Hurricane Ivan 2–24 September 2004*. NOAA National Hurricane Center.

Stolper, D., J.H. List, and E.R. Thieler. 2005. Simulating the evolution of coastal morphology and stratigraphy with a new morphological-behaviour model (GEOMBEST). *Marine Geology* 218(1-4):17-36.

Strozzi, T., P. Teatini, L. Tosi, U. Wegmüller, and C. Werner. 2013. Land subsidence of natural transitional environments by satellite radar interferometry on artificial reflectors. *Journal of Geophysical Research: Earth Surface* 118(2):1177-1191. doi:10.1002/jgrf.20082.

Sturges, W., and B.C. Douglas. 2011. Wind effects on estimates of sea level rise. *Journal of Geophysical Research* 116(C6):C06008. doi:10.1029/2011JC006988.

Stutes, J., J. Cebrian, A.L. Stutes, A. Hunter, and A.A. Corcoran. 2007. Benthic metabolism across a gradient of anthropogenic impact in three shallow coastal lagoons in NW Florida. *Marine Ecology Progress Series* 348:55-70.

Stutz, M.L., and O.H. Pilkey. 2011. Open-ocean barrier islands: Global influence of climatic, oceanographic, and depositional settings. *Journal of Coastal Research* 27(2):207-222. doi:10.2112/09-1190.1.

Sutton-Grier, A.E., K. Wowk, and H. Bamford. 2015. Future of our coasts: The potential for natural and hybrid infrastructure to enhance the resilience of our coastal communities, economies and ecosystems. *Environmental Science and Policy* 51:137-148.

Sweet, W.V., R.E. Kopp, C.P. Weaver, J. Obeysekera, R.M. Horton, E.R. Thieler, and C. Zervas. 2017. *Global and regional sea level rise scenarios for the United States*. NOAA Technical Report NOS CO-OPS 083.

Syvitski, J.P.M., C.J. Vörösmarty, A.J. Kettner, and P. Green. 2005. Impact of humans on the flux of terrestrial sediment to the global coastal ocean. *Science* 308(5720):376-380.

Taylor, N.R., J.L. Irish, I.E. Udoh, M.V. Bilskie, and S.C. Hagen. 2015. Development and uncertainty quantification of hurricane surge response functions for hazard assessment in coastal bays. *Natural Hazards* 77(2):1103-1123.

Texas General Land Office (GLO). 2017. *Texas Coastal Resiliency Master Plan*. March 2017. Available at http://www.glo.texas.gov/coastal-grants/projects/texas-coastal-resiliency-master-plan.html [accessed May 20, 2018].

The Nature Conservancy. 2013. Coastal pipeline erosion control using oyster reefs, Louisiana, USA, Shell Pipeline Company LP. In Green infrastructure case studies: Case studies evaluated by participating companies for creation of the white paper "The case for green infrastructure." Available at https://www.nature.org/about-us/working-with-companies/case-studies-for-green-infrastructure.pdf [accessed May 21, 2018].

The Nature Conservancy. 2018. Gulf of Mexico. Available at https://www.nature.org/ourinitiatives/regions/northamerica/areas/gulfofmexico/index.htm [accessed March 20, 2018].

The Water Institute of the Gulf. 2018. About the Water Institute. Available at https://thewaterinstitute.org/about [accessed April 24, 2018].

Theriot, J.P. 2014. *American energy, imperiled coast: Oil and gas development in Louisiana's wetlands*. Baton Rouge: Louisiana State University Press.

Thomas, C.W., A.B. Murray, A.D. Ashton, M.D. Hurst, A.K.A.P. Barkwith, and M.A. Ellis. 2016. Complex coastlines responding to climate change: Do shoreline shapes reflect present forcing or "remember" the distant past? *Earth Surface Dynamics* 4(4):871-884. doi:10.5194/esurf-4-871-2016.

Thompson, R.R., D.R. Garfin, and R.C. Silver. 2017. Evacuation from natural disasters: A systematic review of the literature. *Risk Analysis* 37(4):812-839.

Timmins, C. 2007. If you cannot take the heat, get out of the cerrado… recovering the equilibrium amenity cost of nonmarginal climate change in Brazil. *Journal of Regional Science* 47(1):1-25.

Törnqvist, T.E., D.J. Wallace, J.E.A. Storms, J. Wallinga, R.L. Van Dam, M. Blaauw, M.S. Derksen, C.J.W. Klerks, C. Meijneken, and E.M.A. Snijders. 2008. Mississippi Delta subsidence primarily caused by compaction of Holocene strata. *Nature Geoscience* 1(3):173-176.

Trenberth, K.E., C.A. Davis, and J. Fasullo. 2007. Water and energy budgets of hurricanes: Case studies of Ivan and Katrina. *Journal of Geophysical Research* 112(D23):D23106.

Tsoar, H. 2005. Sand dunes mobility and stability in relation to climate. Physica A 357(1):50-56.

Turner, B.L., P.A. Matson, J.J. McCarthy, R.W. Corell, L. Christensen, N. Eckley, G.K. Hovelsrud-Broda, J.X. Kasperson, R.E. Kasperson, A. Luers, and M.L. Martello. 2003. Illustrating the coupled human–environment system for vulnerability analysis: three case studies. *Proceedings of the National Academy of Sciences of the United States of America* 100(14):8080-8085.

Turner, J.C., M.A. Hogg, P.J. Oakes, S.D. Reicher, and M.S. Wetherell. 1987. *Rediscovering the social group: A self-categorization theory*. Cambridge, MA: Basil Blackwell.

Turner, R.E., N.N. Rabalais, R.B. Alexander, G. McIsaac, and R.W. Howarth. 2007. Characterization of nutrient, organic carbon, and sediment loads and concentrations from the Mississippi River into the Northern Gulf of Mexico. *Estuaries and Coasts* 30(5):773-790.

Tweel, A.W., and R.E. Turner. 2012a. Landscape-scale analysis of wetland sediment deposition from four tropical cyclone events. *PLOS ONE* 7(11):e50528.

Tweel, A.W., and R.E. Turner. 2012b. Watershed land use and river engineering drive wetland formation and loss in the Mississippi River birdfoot delta. *Limnology and Oceanography* 57(1):18-28.

U.S. Census Bureau. 2018. American FactFinder. Available at https://factfinder.census.gov/faces/nav/jsf/pages/index.xhtml [accessed April 25, 2018].

USACE (U.S. Army Corps of Engineers). 1981. *Galveston's bulwark against the sea, history of the Galveston seawall.* USACE Galveston District. October 1981.

USACE. 2012. Greater New Orleans hurricane and storm damage risk reduction system facts and figures August 2012. Available at http://www.mvn.usace.army.mil/Portals/56/docs/PAO/Brochures/FactsFiguresAugust2012.pdf [accessed at May 20, 2018].

USACE. 2013a. Southwest coastal Louisiana: Draft integrated feasibility report and programmatic environmental impact statement. U.S. Army Corps of Engineers, New Orleans District. December 2013.

USACE 2013b. Summary of the Morganza to the Gulf of Mexico, Louisiana Final Post Authorization Change Report. Available at http://www.mvn.usace.army.mil/Portals/56/docs/PD/Projects/MTG/M2GPACReportMay2013.pdf [accessed May 21, 2018].

USACE. 2015. Sabine Pass to Galveston Bay, Texas coastal storm risk reduction and ecosystem restoration draft integrated feasibility report and environmental impact study draft: Appendix A measure information sheets. Available at http://www.swg.usace.army.mil/Portals/26/docs/Planning/Sabine%20Pass%20to%20Galveston/S2G%20(Keep%20this%20folder)/S2G%20DIFR_EIS_App_A_Part_1.pdf [accessed April 25, 2018].

USACE. 2018. National levee database. Available at http://nld.usace.army.mil [accessed 23 April 2018].

USGS (U.S. Geological Survey). 2017. *Alabama barrier island restoration assessment, Interim report.* USGS, State of Alabama, National Fish and Wildlife Foundation, and USACE, Mobile District, August 2017.

Valiela, I. 2006. *Global coastal change.* Hoboken, NJ: Wiley-Blackwell.

van der Meer, J.W., C. Cooper, M.J. Warner, H. Adams-Morales, and G.J. Steendam. 2008. The success of the hurricane protection around Chevron's refinery at Pascagoula, MS, during Katrina. US PIANC Conference: Gulf Coast Hurricane Preparedness, Response, Recovery and Rebuilding. Mobile, MS. Nov 11–14, 2008.

van Maanen, B., R.J. Nicholls, J. French, A. Barkwith, D. Bonaldo, H. Burningham, A.B. Murray, A. Payo, J. Southerland, G. Thornhill, I.H. Townend, M. van der Wegen, and M.J.A. Walkden. 2016. Simulating mesoscale coastal evolution for decadal coastal management: A new framework integrating multiple, complementary modeling approaches. *Geomorphology* 256:68-80. doi:10.1016/j.geomorph.2015.10.026.

Visser, J.M., S.M. Duke-Sylvester, J. Carter, and W.P. Broussard III. 2013. A computer model to forecast wetland vegetation changes resulting from restoration and protection in coastal Louisiana. *Journal of Coastal Research* 67(sp1):51-59.

Vitousek, S., P.L. Barnard, P. Limber, L. Erikson, and B. Cole. 2017. A model integrating longshore and cross-shore processes for predicting long-term shoreline response to climate change. *Journal of Geophysical Research: Earth Surface* 122(4):782-806.

Wahl, T., and N.G. Plant. 2015. Changes in erosion and flooding risk due to long-term and cyclic oceanographic trends. *Geophysical Research Letters* 42(8):2943-2950.

Walker, N.D. 1996. Satellite assessment of Mississippi River plume variability: Causes and predictability. *Remote Sensing of Environment* 58(1):21-35.

Walker, W., M. Haasnoot, and J. Kwakkel. 2013. Adapt or perish: A review of planning approaches for adaptation under deep uncertainty. *Sustainability* 5(3):955-979.

Wallace, D.J., and J.B. Anderson. 2013. Unprecedented erosion of the upper Texas Coast: Response to accelerated sea level rise and hurricane impacts. *Geological Society of America Bulletin* 125: 7286740.

Wallace, D.J., J.B. Anderson, and R. Fernandez. 2010. Transgressive ravinement versus depth of closure: A geological perspective from the upper Texas coast. *Journal of Coastal Research* 26(6):1057-1067.

Walters, D., L.J. Moore, O. Durán, S. Fagherazzi, and G. Mariotti. 2014. Interactions between barrier islands and backbarrier marshes affect island system response to sea level rise: Insights from a coupled model. *Journal of Geophysical Research: Earth Surface* 119(9):2013-2031. doi:10.1002/2014JF003091.

Wamsley, T.V., M.A. Cialone, J.M. Smith, B.A. Ebersole, and A.S. Grzegorzewski. 2009. Influence of landscape restoration and degradation on storm surge and waves in southern Louisiana. *Natural Hazards* 51(1):207-224.

Wamsley, T.V., M.A. Cialone, J.M. Smith, J.H. Atkinson, and J.D. Rosati. 2010. The potential of wetlands in reducing storm surge. *Ocean Engineering* 37(1):59-68. doi:10.1016/j.oceaneng.2009.07.018.

Webb, E.L., D.A. Friess, K.W. Krauss, D.R. Cahoon, G.R. Guntenspergen, and J. Phelps. 2013. A global standard for monitoring coastal wetland vulnerability to accelerated sea-level rise. *Nature Climate Change* 3(5):458-465. doi:10.1038/nclimate1756.

Weber, E.U., and C.K. Hsee. 1998. Cross-cultural differences in risk perception, but cross-cultural similarities in attitudes towards perceived risk. *Management Science* 44(9):1205-1217.

Weerts, D.J., and L.R. Sandmann. 2008. Building a two-way street: Challenges and opportunities for community engagement at research universities. *The Review of Higher Education* 32(1):73-106.

Wellner, R., R. Beaubouef, J. Van Wagoner, H. Roberts, and T. Sun. 2005. Jet-plume depositional bodies—The primary building blocks of Wax Lake Delta. *Gulf Coast Association of Geological Societies Transactions* 55:867-909.

Werner, B.T. and D.E. McNamara. 2007. Dynamics of coupled human-landscape systems. *Geomorphology* 91(3-4):393-407.

Weston, N.B. 2014. Declining sediments and rising seas: An unfortunate convergence for tidal wetlands. *Estuaries and Coasts* 37(1):1-23.

White, E.D., E. Meselhe, A. McCorquodale, B. Couvillion, Z. Dong, S.M. Duke-Sylvester, and Y. Wang. 2017. *2017 Coastal Master Plan: Attachment C2-22: Integrated Compartment Model (ICM) development. Version final.* Baton Rouge, Louisiana: Coastal Protection and Restoration Authority.

White, W.A., and T.A. Tremblay. 1995. Submergence of wetlands as a result of human-induced subsidence and faulting along the upper Texas Gulf Coast. Coastal Education & Research Foundation, Inc. *Journal of Coastal Research* 11(3):788-807.

Whyte, W.F.E. 1991. *Participatory action research.* Newbury Park, CA: Sage Publications, Inc.

Wiberg, P.L., J.A. Carr, I. Safak, and A. Anutaliya. 2015. Quantifying the distribution and influence of non-uniform bed properties in shallow coastal bays. *Limnology and Oceanography: Methods* 13(12):746-762.

Wicker, K.M., R.E. Emmer, D. Roberts, and J. van Beek. 1989. *Pipelines, navigation channels, and facilities in sensitive coastal habitats, an analysis of outer continental shelf impacts, coastal Gulf of Mexico. Volume I: technical narrative.* OCS Report/ MMS 89-0051. U.S. Department of the Interior—Minerals Management Service. Gulf of Mexico OCS Regional Office. New Orleans, LA.

Wilbanks, T., and S. Fernandez (Coordinating Lead Authors). 2013. *Climate change and infrastructure, urban systems, and vulnerabilities.* Technical report for the U.S. Department of Energy in Support of the National Climate Assessment. Washington, DC: Island Press.

Williams, Z.C., D.E. McNamara, M.D. Smith, A.B. Murray, and S. Gopalakrishnan. 2013. Coupled economic-coastline modeling with suckers and free riders. *Journal of Geophysical Research: Earth Surface* 118(2):887-899. doi:10.1002/jgrf.20066.

Wilson, G.W., H.T. zkan-Haller, R.A. Holman, M.C. Haller, D.A. Honegger, and C.C. Chickadel. 2014. Surf zone bathymetry and circulation predictions via data assimilation of remote sensing observations. *Journal of Geophysical Research: Oceans* 119(3):1993-2016. doi:10.1002/2013JC009213.

Wilson, S.G., and T.R. Fischetti. 2010. *Coastline population trends in the United States: 1960 to 2008. Population estimates and projections.* Current Population Reports. U.S. Census Bureau. https://www.census.gov/prod/2010pubs/p25-1139.pdf [accessed May 15, 2018].

Winslow, A.G., and W.W. Doyel. 1954. Land-surface subsidence and its relation to the withdrawal of ground water in the Houston–Galveston region, Texas. *Economic Geology* 49(4):413-422. doi:10.2113/gsecongeo.49.4.413.

Withey, L., K. Seto, D.J. McCauley, K.J. Fiorella, R.A. Marsh, B. Abrahms, T.A. Nuñez, C.D. Golden, and J.S. Brashares. 2014. Fauna in decline—Response. *Science* 346(6211):819-820.

Wolstencroft, M., Z. Shen, T.E. Törnqvist, G.A. Milne, and M. Kulp. 2014. Understanding subsidence in the Mississippi Delta region due to sediment, ice, and ocean loading: Insights from geophysical modeling. *Journal of Geophysical Research: Solid Earth* 119(4):3838-3856. doi:10.1002/2013jb010928.

Wong-Parodi, G., and I. Feygina. Under review. How context, attitudes, and psychosocial factors influences (mal) adaptive responses to natural disasters: The case of Hurricane Matthew. *Weather, Climate, and Society.*

Wong-Parodi, G., H. Dowlatabadi, T. McDaniels, and I. Ray. 2011. Influencing attitudes toward carbon capture and sequestration: A social marketing approach. *Environmental Science and Technology* 45(16):6743-6751.

Wong-Parodi, G., W.B. de Bruin, and C. Canfield. 2013. Effects of simplifying outreach materials for energy conservation programs that target low-income consumers. *Energy Policy* 62:1157-1164.

Wong-Parodi, G., B. Fischhoff, and B. Strauss. 2017. Plans and prospects for coastal flooding in four communities affected by Sandy. *Weather, Climate, and Society* 9(2):183-200. doi:10.1175/WCAS-D-16-0042.1.

Wong-Parodi, G., B. Fischhoff, and B. Strauss. 2018. Effect of risk and protective decision aids on flood preparation in vulnerable communities. *Weather, Climate, and Society* 10:401-417. doi:10.1175/WCAS-D-17-0069.1.

Woodbury, H.O., I.B. Murray, Jr., P.J. Pickford, and W.H. Akers. 1973. Pliocene and Pleistocene depocenters, outer continental shelf, Louisiana and Texas. *American Association of Petroleum Geologists Bulletin* 57:2428-2439.

Woodruff, J.D., J.L. Irish and S.J. Camargo. 2013. Coastal flooding by tropical cyclones and sea level rise. *Nature* 504(7478):44-52.

Wright, D.B., T.R. Knutson, and J.A. Smith. 2015. Regional climate model projections of rainfall from U.S. landfalling tropical cyclones. *Climate Dynamics* 45(11-12):3365-3379.

Yan, X., R. Zhang, and T.R. Knutson. 2017. The role of Atlantic overturning circulation in the recent decline of Atlantic major hurricane frequency. *Nature Communications* 8(1):1695.

Yeager, K.M., C.A. Brunner, M.A. Kulp, D. Fischer, R.A. Feagin, K.J. Schindler, J. Prouhet, and G. Bera. 2012. Significance of active growth faulting on marsh accretion processes in the lower Pearl River, Louisiana. *Geomorphology* 153-154:127-143. doi:10.1016/j.geomorph.2012.02.018.

Yu, J., and G. Wang. 2016. GPS-derived ground deformation (2005–2014) within the Gulf of Mexico region referred to a stable Gulf of Mexico reference frame. *Natural Hazards and Earth System Sciences* 16(7):1583-1602. doi:10.5194/nhess-16-1583-2016.

Zarnetske, P.L., S.D. Hacker, E.W. Seabloom, P. Ruggiero, J.R. Killian, T.B. Maddux, and D. Cox. 2012. Biophysical feedback mediates effects of invasive grasses on coastal dune shape. *Ecology* 93:1439-1450. doi:10.1890/11-1112.1.

Committee and Staff Biographical Sketches

TUBA ÖZKAN-HALLER (*Committee Chair*) is a Professor and Associate Dean for Research and Faculty Advancement at the College of Earth, Ocean, and Atmospheric Sciences and School of Civil and Construction Engineering at Oregon State University. Her research program focuses on the use of numerical, field, laboratory, and analytical approaches to predict water motion and bathymetric change in bays, inlets, and the continental shelf. She has developed models to predict surf zone wave fields, conducted research to understand the effects of wave energy converters on the nearshore wave field, and investigated the mechanisms that control nearshore rip currents. Dr. Özkan-Haller has more recently been involved in studies dealing with oxygen consumption in the coastal ocean. She is a member of the Ocean Studies Board of the National Academies of Sciences, Engineering, and Medicine and served on the Committee on an Evaluation of the U.S. Department of Energy's Marine and Hydrokinetic Resource Assessments for the National Academies. Dr. Özkan-Haller is the recipient of the Office of Naval Research Young Investigator Award and the Outstanding Faculty Member Award at the University of Michigan. She holds a B.S. in Civil Engineering from Boğaziçi University in Istanbul, Turkey, and an M.C.E. and a Ph.D. in Civil Engineering from the University of Delaware.

GREGORY A. CARTER is a Professor of Geography in the Department of Geography and Geology and a Chief Scientist of the Gulf Coast Geospatial Center at the University of Southern Mississippi. He and his students study the ecological communities of barrier islands and wetlands in the northern Gulf of Mexico. They apply present-day and historical remote sensing, as well as field sampling techniques, to investigate relationships among ecosystems, elevation, sea level rise, and the impact of tropical storms. Dr. Carter received B.S. and M.S. degrees in Botany from Auburn University and a Ph.D. in Botany from the University of Wyoming.

JUST CEBRIAN is a Professor in the Department of Marine Sciences at the University of South Alabama and a Senior Marine Scientist at the Dauphin Island Sea Lab in Alabama. His research focuses on the impacts that humans have on the functioning of coastal ecosystems to better understand coastal resiliency and help generate cost-effective management policies. His field and experimental work follows a community-integrated approach and covers the main communities of coastal ecosystems, including phytoplankton communities, sediment flats inhabited by benthic microalgae, macroalgal beds, seagrass meadows, and marshes. Dr. Cebrian also studies how coastal ecosystems compare with terrestrial ecosystems using a number of functional metrics, including trophic processes and energy flows. Dr. Cebrian serves on the editorial boards of the journals *Marine Ecology Progress Series*, *Estuaries and Coasts*, *PLOS ONE*, and *Gulf and Caribbean Research*. He obtained a B.A. in Biology from the University of Barcelona, Spain, and an M.S. in Oceanology from the University of Perpignan in France. He received his Ph.D. in Marine Sciences at the University of Politecnica Catalunya in Barcelona, Spain.

ROBERT (TONY) A. DALRYMPLE (NAE) is the Willard and Lillian Hackerman Professor Emeritus of Civil Engineering at Johns Hopkins University and a Distinguished Professor of Coastal Engineering at Northwestern University. His major research interests are in the areas of coastal engineering, wave mechanics, fluid mechanics, littoral processes, and tidal inlets. His research currently explores water wave modeling, including tsunamis and their impacts on shorelines. Dr. Dalrymple was elected to the National Academy of Engineering in 2006. He chaired the National Academy of Sciences' Committee on the Review of the Louisiana Coastal Protection and Restoration Program and the Committee on Sea Level Rise in California, Oregon, and Washington. Dr. Dalrymple received his A.B. degree in Engineering Sciences from Dartmouth College, his M.S. degree in Ocean Engineering from the University of Hawaii, and his Ph.D. degree in Civil and Coastal Engineering from the University of Florida.

JORDAN R. FISCHBACH is a Senior Policy Researcher at the RAND Corporation, Co-Director of the Water and Climate Resilience Center, and an affiliate faculty member at the Pardee RAND Graduate School. Dr. Fischbach has expertise in risk analysis, exploratory simulation modeling, and Robust Decision Making, a method designed to better manage deep uncertainty and develop robust and adaptive plans through quantitative scenario analysis. He currently serves as Co-Program Manager, Science and Analysis, and Co-Investigator for the National Oceanic and Atmospheric Administration (NOAA) Mid-Atlantic Regional Integrated Sciences and Assessments Center, and is leading an assessment of damage and needs after Hurricane Maria to support Puerto Rico's recovery planning. Dr. Fischbach recently served as the principal investigator for the flood risk and damage assessment supporting the State of Louisiana's *2017 Coastal Master Plan*, and co-led an investigation of future climate impacts and coastal resilience options in Queens after Hurricane Sandy. Dr. Fischbach earned a B.A. in History from Columbia University and a Ph.D. in Policy Analysis from the Pardee RAND Graduate School, where he was awarded the Herbert Goldhamer Memorial Award.

JENNIFER L. IRISH is a Professor of Coastal Engineering in the Department of Civil and Environmental Engineering at Virginia Tech. Previously, she was the Coastal Engineering Regional Technical Specialist at the U.S. Army Corps of Engineers North Atlantic Division and New York District, and a Research Coastal Engineer at the U.S. Army Coastal and Hydraulic Laboratory. Dr. Irish's research involves the physical impacts of coastal hazards, including storm surge, tsunami inundation, and storm-induced erosion; coastal hazard probability and risk assessment; impacts of climate change and sea level rise at the coast; and the role of natural and nature-based features in coastal hazard mitigation, including wetlands, coastal forest, dunes, and beaches. She is a Fellow of the American Society of Civil Engineers (ASCE) and has received numerous awards, including the Department of the Army's Superior Civilian Service Award. Dr. Irish serves on multiple editorial boards, including the journal *Coastal Engineering*. She was formerly Chair of ASCE's Committee on Technical Advancement and Secretary of ASCE's Coasts, Oceans, Ports, and Rivers Institute. Dr. Irish received her B.S. and M.S. in Civil Engineering from Lehigh University in Pennsylvania and her Ph.D. in Civil Engineering from the University of Delaware.

ALEXANDER S. KOLKER is an Associate Professor in the Louisiana Universities Marine Consortium and teaches in the Department of Earth and Environmental Science at Tulane University. Dr. Kolker's research lab investigates interactions among sedimentary, hydrological, and oceanographic processes in the coastal zone and the ways these processes affect and are affected by the morphology of coasts and wetlands. Dr. Kolker's work spans the natural and anthropogenic processes that govern coastal systems and the role of atmospheric processes on short-term sea level dynamics, and works to understand how climate and other human activities influence coasts and wetlands. His current projects include the development of a subsidence map of the Louisiana coast; an examination of the influence of the Mississippi River and its delta on the oceanography and ecology of the Gulf of Mexico; investigations into natural analogues for coastal restoration in Louisiana; and studies investigating the pathways and processes associated with groundwater discharge in the Mississippi River Delta. He holds a B.A. in Biology from the University of California, Santa Cruz; an M.A. in Biological Sciences; and a Ph.D. in Marine and Atmospheric Science from the State University of New York, Stony Brook.

SHUBHRA MISRA worked as a marine and coastal subject-matter expert for 6 years at Chevron Energy Technology Company in Houston on Chevron's oil, gas, and petrochemical infrastructure projects globally. Prior to Chevron, Dr. Misra worked at several marine and coastal infrastructure design and engineering consulting firms in New York and Houston as a Coastal Engineer. Dr. Misra is a Professional Engineer (Civil-Water Resources) registered in Texas. His primary experience is with multidisciplinary coastal and marine projects with a focus on site selection, conceptual and detailed design, constructability and operation of marine and coastal infrastructure (fixed and floating); intakes and outfalls; dredging and reclamation; numerical and physical laboratory modeling of coastal processes (e.g., waves, water levels, currents, sediments) and vessel motions; meteorological-ocean studies; risk assessments; computational fluid dynamics modeling including wave-structure interactions; landslide tsunamis; shipping waterway suitability assessments; and marine environmental impact assessments. He received his M.S. and Ph.D. degrees from the University of Delaware in Civil and Environmental Engineering (coastal engineering), followed by a brief stint as a postdoctoral researcher at the College of Marine Studies, University of Delaware, conducting laboratory experiments on air–sea interactions.

LAURA J. MOORE is an Associate Professor and Director of the Coastal Environmental Change Laboratory in the Department of Geological Sciences and the Curriculum for the Environment and Ecology at the University of North Carolina at Chapel Hill. Her interdisciplinary research program focuses on the response of low-lying coastal environments to climate change. Recent and ongoing work relies on the merging of observational and numerical approaches to investigate barrier island response to sea level rise; coastal foredune dynamics; couplings among barriers, back-barrier marshes, and bays; large-scale coastline response to changing wave climate; two-way couplings between human activities and natural processes that affect coastline evolution; and novel arts-based approaches to climate change education. Her research involves collaboration with ecologists, social scientists, and performance artists. Dr. Moore has been a principal investigator at the Virginia Coast Reserve Long-Term Ecological Research site since 2008; is the co-editor of *Barrier Dynamics and Response to Changing Climate*, published in January 2008 by Springer; and is a recipient of the W.M. Keck Foundation Fellowship in Natural Sciences, among others. Dr. Moore has a B.A. in Geology from Colgate University and a Ph.D. in Earth Sciences from the University of California, Santa Cruz.

MARTIN D. SMITH is the George M. Woodwell Distinguished Professor of Environmental Economics in the Nicholas School of the Environment at Duke University and also has an appointment in the Department of Economics at Duke University. Dr. Smith's research focuses on the economics of the oceans, including fisheries, marine ecosystems, seafood markets, and coastal climate adaptation. He has written on a range of policy-relevant topics, including economics of marine reserves, seasonal closures in fisheries, ecosystem-based management, nutrient pollution, the global seafood trade, organic agriculture, and coastal responses to climate change. He is best known for identifying unintended consequences of marine and coastal policies that ignore human behavioral feedbacks. Dr. Smith's methodological interests span micro-econometrics, optimal control theory, time series analysis, and numerical modeling of coupled human-natural systems. He serves on the Ocean Studies Board of the National Academies of Sciences, Engineering, and Medicine; is an Aldo Leopold Leadership Fellow; and was selected for the Quality of Research Discovery Award from the Agricultural and Applied Economics Association. He is former Editor-in-Chief of *Marine Resource Economics* and has a B.A. in Public Policy from Stanford University and a Ph.D. in Agricultural and Resource Economics from the University of California, Davis.

TORBJÖRN E. TÖRNQVIST is the Vokes Geology Professor in the Department of Earth and Environmental Sciences at Tulane University. Dr. Törnqvist is a Quaternary scientist who studies sea level change, coastal subsidence, delta evolution, and paleoclimatology. More specifically, he examines the sedimentary record of the Louisiana coast to investigate sea level change over timescales ranging from the past decade to the past 10,000 years. Sea-level records provide insights on a variety of issues, including rates of ice sheet melt during past warm periods, as well as subsidence mechanisms, rates, and their spatial patterns. Dr. Törnqvist's research group also examines the response of coastal and deltaic environments to accelerated rates of sea level rise, including studies that seek to assess how deltaic processes can be harnessed to benefit coastal restoration. He received his M.S. and Ph.D. degrees in Physical Geography from Utrecht University.

GABRIELLE WONG-PARODI is an Assistant Research Professor in the Department of Engineering and Public Policy at Carnegie Mellon University (CMU). Her research focuses on applying behavioral decision research methods to promote environmental sustainability and community resiliency along the coastline. Dr. Wong-Parodi uses behavioral science approaches to create evidence-based strategies for informed decision making across a range of domain areas. These include energy resources, climate change adaptation and mitigation, and emerging technologies, such as autonomous vehicles and unconventional shale gas development. She was an invited speaker at the Sackler Colloquia on the Science of Science Communication at the National Academy of Sciences. Dr. Wong-Parodi is a faculty affiliate at Lawrence Berkeley National Laboratory and is the social science research liaison for CMU at the Climate Advocacy Lab. Dr. Wong-Parodi received her B.A. in Psychology at the University of California, Berkeley, and her M.A. and Ph.D. in Risk Perceptions and Communication from the University of California, Berkeley.

National Academies Staff

DEBORAH GLICKSON (*Study Director*) is a Senior Program Officer with the Board on Earth Sciences and Resources at the National Academies of Sciences, Engineering, and Medicine. She received an M.S. in Geology from Vanderbilt University and a Ph.D. in Oceanography from the University of Washington. Her doctoral research focused on magmatic and tectonic contributions to mid-ocean ridge evolution and hydrothermal activity at the Endeavour Segment of the Juan de Fuca Ridge. After finishing her Ph.D., she participated in the Dean John A. Knauss Marine Policy Fellowship and worked on coastal and ocean policy and legislation in the U.S. Senate. Prior to her Ph.D. work, she was a Research Associate in Physical Oceanography at Woods Hole Oceanographic Institution. Since joining the National Academies staff in 2008, she has worked on several ocean and Earth science studies, including such topics as scientific ocean drilling, critical ocean science research needs and infrastructure, the academic research fleet, marine hydrokinetic energy, methane hydrates, coal mining and human health, and geoscience education.

HEATHER KREIDLER is an Associate Program Officer for the National Academies' Board on Environmental Change and Society and the Board on Human-Systems Integration. She joined the National Academies in 2008 and has worked on wide-ranging topics including public health, nutrition, and dietary guidance, and issues facing children, youth, and families. Her current projects examine and advance the social and behavioral sciences at the intersection of human activity and global environmental change and issues concerning the relationship of individuals and organizations to technology and the environment. Ms. Kreidler received a B.S. in Business Management from Kutztown University in Pennsylvania and an M.S. in Environmental Science and Policy from George Mason University.

COURTNEY DEVANE is the Administrative Coordinator for the National Academies' Board on Earth Sciences and Resources and the Water Science and Technology Board. She received an A.A. in Graphic Design from Pittsburgh Technical Institute. She joined the National Academies staff in 2004 and has worked across the institution on a wide variety of projects and subjects—most notably with the National Academy of Engineering, the National Academy of Sciences, the Naval Studies Board, the Board on Radiation Effects Research, and the Nuclear and Radiation Studies Board.

JAMES HEISS was a Postdoctoral Fellow with the National Academies' Ocean Studies Board. He graduated with an M.S. and a Ph.D. in Geology from the University of Delaware. His research involved understanding the role of waves and tides on controlling the temporal and spatial variability of groundwater flow and saltwater-freshwater mixing in coastal aquifers. He has also researched linkages between groundwater hydrology and biogeochemical cycling in beach groundwater systems. At the National Academies, he worked on studies dealing with the use of chemical dispersants in oil spill response and the NASA Decadal Survey for Earth Science and Applications from Space.

People Who Provided Input to the Committee

Meeting 1: Washington, DC
May 17, 2017

Tom Drake, Ocean, Atmosphere and Space Research, Office of Naval Research
Nicole Elko, American Shore and Beach Preservation Association
David Kidwell, National Centers for Coastal Ocean Science, National Oceanic and Atmospheric Administration
Tucker Mahoney, Federal Emergency Management Agency
Julie Dean Rosati, Coastal and Hydraulics Laboratory, U.S. Army Corps of Engineers
Hilary Stockdon, Coastal and Marine Geology Program, U.S. Geological Survey
George Voulgaris, Physical Oceanography Program, National Science Foundation

Meeting 2: Houston, Texas
July 18–19, 2017

Phil Berke, Texas A&M University
Carl Bernier, Rice University
Cas Bridge, Chevron
Noreen Clancy, RAND Corporation
Don Danmeier, Chevron
Carl Ferraro, Alabama Department of Conservation and Natural Resources
Angelina Freeman, Louisiana Coastal Protection and Restoration Authority
Maria Lemos, University of Michigan
Ryan Moyer, Florida Department of Environmental Protection
Ray Newby, Texas General Land Office
Natalie Peyronnin, Environmental Defense Fund

George Ramseur, Mississippi Department of Marine Resources
Denise Reed, The Water Institute of the Gulf
Ariana Sutton-Grier, University of Maryland
LaDon Swann, Mississippi-Alabama Sea Grant Consortium
Jason Theriot, historian and author
Elizabeth Vargas, Texas General Land Office

Meeting 3: New Orleans, Louisiana
September 18, 2017

Damarys Acevedo-Mackey, U.S. Army Engineer Research and Development Center
Mead Allison, The Water Institute of the Gulf
Joe Calantoni, Naval Research Laboratory
Brady Couvillion, U.S. Geological Survey
David Dismukes, Louisiana State University
John Alex McCorquodale, University of New Orleans
Ehab Meselhe, The Water Institute of the Gulf
Hugh Roberts, ARCADIS
Dano Roelvink, IHE Delft Institute for Water Education
Jenneke Visser, University of Louisiana at Lafayette
Ian Voparil, Shell Deepwater Gulf of Mexico
Ty Wamsley, U.S. Army Corps of Engineers Mississippi Valley Division
Eric White, The Water Institute of the Gulf

Webinar
November 1, 2017

Brad Murray, Duke University

Meeting 4: St. Petersburg, Florida
November 15, 2017

Craig Colten, Louisiana State University
Mathew Hauer, University of Georgia
Kelli Levy, Environmental Management Division Director, Pinellas County
Shahzaad Mohammed, Cheniere LNG
Jim Schock, Florida Building Commission
Sherri Swanson, Environmental Project Manager, HDR Engineering, Inc.